SUBTERRANEAN MATTERS

ELEMENTS *A series edited by*
Stacy Alaimo and Nicole Starosielski

SUBTERRANEAN MATTERS

COOPERATIVE MINING AND RESOURCE
NATIONALISM IN PLURINATIONAL BOLIVIA

ANDREA MARSTON

DUKE UNIVERSITY PRESS   DURHAM AND LONDON   2024

Project Editor: Michael Trudeau
Typeset in Chaparral Pro and Knockout by BW&A Books, Inc.

Library of Congress Cataloging-in-Publication Data
Names: Marston, Andrea [date] author.
Title: Subterranean matters : cooperative mining and resource
nationalism in plurinational Bolivia / Andrea Marston.
Other titles: Elements (Duke University Press)
Description: Durham : Duke University Press, 2024. | Series: Elements |
Includes bibliographical references and index.
Identifiers: LCCN 2023015287 (print)
LCCN 2023015288 (ebook)
ISBN 9781478025634 (paperback)
ISBN 9781478020899 (hardcover)
ISBN 9781478027768 (ebook)
Subjects: LCSH: Mines and mineral resources—Political aspects—
Bolivia. | Mineral industries—Political aspects—Bolivia. | Mines
and mineral resources—Bolivia. | Mineral industries—Bolivia. |
Cooperative societies—Bolivia. | Tin industry—Bolivia. | BISAC:
POLITICAL SCIENCE / Labor & Industrial Relations | HISTORY /
Latin America / South America
Classification: LCC HD9506.B52 M377 2024 (print) |
LCC HD9506.B52 (ebook) | DDC 338.20984—DC23/ENG/20231024
LC record available at https://lccn.loc.gov/2023015287
LC ebook record available at https://lccn.loc.gov/2023015288

Cover art: Cooperative miners working underground.
Llallagua, Bolivia, 2017. Photo by the author.

For all my families

Abbreviation	Spanish	English
CEDLA	Centro de Estudios para el Desarrollo Laboral y Agrario	Research Center for Labor and Agrarian Development
CIDOB	Confederación de Pueblos Indígenas de Bolivia	Confederation of Indigenous Peoples of Bolivia
COB	Central Obrera Boliviana	Bolivian Workers' Central
COMIBOL	Corporación Minera de Bolivia	Mining Corporation of Bolivia
CONAMAQ	Consejo Nacional de Ayllus y Markas del Qullasuyu	National Council of Ayllus and Markas of Qullasuyu
CPE	Constitución Política del Estado	Political Constitution of the State
CSUTCB	Confederación Sindical Única de Trabajadores Campesinos de Bolivia	Unified Syndical Confederation of Peasant Workers of Bolivia
DENAGEO	Departamento Nacional de Geología	National Department of Geology
DS	Decreto Supremo	Supreme Decree
FENCOMIN	Federación Nacional de Cooperativas Mineras de Bolivia	National Federation of Mining Cooperatives of Bolivia
FERECOMINORPO	Federación Regional de Cooperativas Mineras del Norte de Potosí	Regional Federation of Mining Cooperatives of Northern Potosí
FONDIOC	Fondo de Desarrollo para los Pueblos Indígenas Originarios y Comunidades Campesinas	Fund for the Development of Original Indigenous Peoples and Peasant Communities

Abbreviation	Spanish	English
FPS	Formación Política Sindical	Political Syndicalist Formation
FSTMB	Federación Sindical de Trabajadores Mineros de Bolivia	Syndical Federation of Mine Workers of Bolivia
GEOBOL	Servicio Geológico de Bolivia	Geological Service of Bolivia
IDH	Impuesto Directo a los Hidrocarburos	Direct Tax on Hydrocarbons
ISI		import substitution industrialization
ITC		International Tin Committee
MAS	Movimiento al Socialismo	Movement toward Socialism
MNR	Movimiento Nacional Revolucionario	National Revolutionary Movement
NGO		nongovernmental organization
SERGEOMIN	Servicio Nacional de Geología y Minería	National Geology and Mining Service
THOA	Taller de Historia Oral Andina	Andean Oral History Workshop
TIPNIS	Territorio Indígena y Parque Nacional Isiboro Sécure	Isiboro Secure Indigenous Territory and National Park
UNSXX	Universidad Nacional Siglo XX	National University "Siglo XX" (Twentieth Century)

In March 2016 I rested on a park bench in Sopocachi, a trendy neighborhood of La Paz known for its upscale cafés, bars, and urban activists. I opened a book I had just purchased: the first of five volumes of *Historia del movimiento obrero boliviano 1933–1952* (History of the Bolivian workers' movement, 1933–1952; 1980), by Guillermo Lora, a Trotskyist historian who came of age in the Bolivian tin belt in the 1940s. Several academic acquaintances had made it clear that his books were required reading for anyone hoping to study mining in Bolivia, and I was pleased to have finally found the whole series tucked away in an alley of book vendors.

I had read less than two pages, however, before the man on the bench next to me grew curious about why a gringa was reading Bolivian labor history. One of the older middle-class *paceños* (people from La Paz) who visit the park every afternoon to warm themselves in the high-altitude sun, this man had a distinctly grandfatherly presence. He leaned forward as he spoke, eager to tell me that his father had been an engineer in the tin mines and that he himself had been radicalized listening to union leaders' speeches as a young man. He asked me if I, too, was learning from the miners' unions. I smiled as I shook my head, explaining that although I was interested in the workers' movement, the focus of my research was the *cooperativas mineras* (mining cooperatives). These collectives of small-scale, independent miners not only incorporated far more members than the miners' unions but also seemed to have an outsized impact on Bolivia's political landscape. I wanted to understand how mining cooperatives had acquired this power, as well as their broader role within Bolivia's extractive economy.

As I spoke, the change on the man's face was dramatic. He leaned away, the edges of his mustache bristling. "Why would you want to study those *cabrones* [assholes]?" he hissed. "They're not really cooperatives, you know. They're thieves—they take what belongs to everyone. They are thieves of *patria* [nation]." He rotated his whole body away, suddenly very interested in the flocking pigeons, and refused to engage me further.

Although the vehemence of this exchange was unusual, interactions of this variety happened to me all the time. By and large, Bolivians are not very fond of cooperative miners. The reasons for this are mixed, but the themes of greed and theft are recurrent. Despite the collectivism implied by their name, cooperative miners have a reputation for extreme individualism. This notoriety derives in part from how they operate internally. Although mining cooperatives have collective mining contracts, they labor alone or in small groups, rarely sharing profits or supplies across the whole cooperative. But their reputation also derives from their relationship to the nation-state, as mediated by the subterranean. The mineralized resources that mining cooperatives exploit are technically the property of all Bolivians, held in trust by the state. Not only do *cooperativistas* (cooperative miners) wrangle personal wealth from the earth, but, because they are legally considered not-for-profit organizations, they are exempt from paying income tax. Finally, the general sense that cooperative miners are quietly skimming from a shared inheritance is exacerbated by accounts of their historical origins, in which mining cooperatives are framed as reactionary groups that emerged in the 1980s from the neoliberalized ashes of the country's once famously progressive miners' unions. From the perspective of most left-leaning Bolivians, mining cooperatives were created by neoliberal policies and remain the purest expression of neoliberal capitalism: selfish, unregulated, and insatiable.

This framing, however, does not align neatly with cooperative miners' actions on the national political stage. Despite being figured as politically reactionary, mining cooperatives were in many ways central to the rise of the left-leaning president Evo Morales, as they are quick to remind anyone who asks. Evo, as he is fondly known by most Bolivians, held power from 2006 to 2019. Not only was he the country's first Indigenous-identifying president, but he is also credited with ushering in a new era of leftist nationalism in Bolivia. With the introduction of a landmark constitution in 2009, the Republic of Bolivia was transformed into the Plurinational State of Bolivia, a seemingly nominal change that nevertheless shook the liberal equation of one (singular) state with one (singular) nation. Although the demand for plurinationalism arose within Indigenous nations and federations, and although Evo's election was hailed internationally as a victory for Indigenous peoples, cooperative miners and other less clearly progressive segments of society also actively participated in both his electoral campaign and the constitutional assembly that he subsequently organized. Because of their contributions, mining cooperatives are now constitutionally recognized as "productive actors" in the mining sector, alongside state-owned and private mining companies. Cooperative miners joined Evo's party, the MAS (Movimiento al Socialismo; Movement toward Socialism), in droves, and over the next thirteen years, they elected dozens of cooperative miners to serve as parliamentary deputies or senators on MAS tickets. From these vantage points, cooperative miners participated in drafting (and blocking) legislation that affects not only the mining sector but all aspects of society. At the time that I was reading Guillermo Lora in Sopocachi, cooperative miners remained among Evo's most ardent supporters.

What were these supposed "thieves of *patria*" doing in the heart of a leftist nation-building project? How did subterranean resources become such a defining feature of Bolivian nationhood, and how did the activity of mining become both nationalist and antinationalist? How were thieves of the subterranean imperative to the rise of Bolivia's leftist government, when they seemed to contradict the administration's insistence on economic redistribution and environmentally sustainable development? These questions had already been circling my brain for several years, but they seemed to congeal after this conversation on the park bench. They continued to guide my fieldwork for the subsequent year and a half, and they remain the overarching concerns of this book.

These questions are no less salient today than they were in 2016, despite the dramatic political changes that have since occurred in Bolivia and across Latin America. At that time, Latin America was at the height of what analysts frequently described as a "pink tide" or "new left" political shift. Since the early 2000s, Latin American countries had been electing administrations that embraced different sets of left-wing policies, often while retaining some elements of neoliberal "good governance" such as fiscal conservativism (hence a pink rather than red tide). In the spirit of critical encouragement, progressive scholars within and beyond Latin America set to work documenting continuities and ruptures between these new administrations and their broadly neoliberal predecessors, often pointing to Evo's Bolivia as the quintessential example of a new left government. This was the conversation I imagined myself joining when I conceived this project in 2012. As other scholars had observed, one of the primary challenges facing Latin America's leftist administrations was how to unite the old left, which had been dominated by trade unionists prior to neoliberalization, and the multiple faces of the new left, such as Indigenous federations, urban social movements, and student associations. This dilemma seemed to manifest *within* the mining cooperatives, for whom both trade unionism and Indigenous political organizing—not to mention the adverse effects of neoliberal economics—figured large in collective memory and daily life. I expected that the specificities of their story would resonate with the challenges facing the Latin American left more broadly.

Since then, the political map of Latin America seems to have flipped twice. Although right-wing governments gained a brief regional hegemony, the pink tide appears newly emergent: in 2018 left-leaning Andrés Manuel López Obrador ended a long series of right-wing Mexican presidents; in 2021 former teacher and union leader Pedro Castillo became president of Peru; and in the watershed year 2022, former student leader Gabriel Boric took over the Chilean presidency, Gustavo Petro became Colombia's first left-wing president ever, and trade unionist Luiz Inácio Lula da Silva was reelected in Brazil, replacing notoriously reactionary Jair Bolsonaro. While the remarkable Evo Morales was ousted from office in 2019 and replaced by ultraright interim president Jeanine Áñez, his party returned to power in 2020, now under the leadership of President Luis Arce. In this context the challenges that faced the Latin American left in the early 2000s are once again at the forefront of much political debate. Moreover, the narrow margin of many leftist victories—not to mention the rapid rate of presidential turnover, with some countries flipping between reactionary and progressive poles with each successive election—underscores the need to attend to the continuities that stitch together apparently divergent political programs. How was the new left built on top of the sedimentary remains of previous eras, and what major fault lines appeared through this process?

Subterranean Matters suggests that a shared feature of left-wing and right-wing politics in Bolivia is a persistent commitment to nation and nationalism. A populist sentiment, nationalism is as readily articulated with progressive as reactionary politics, but it always involves drawing lines on a map and lines within a body politic. Such nationalist lines are conspicuous in right-wing policies such as scaled-back minority rights, tightened immigration policies, and infrastructure such as prisons and border walls, but they are equally at work in many leftist appeals to *pueblo* (people) and *patria* (nation). Nationalism is key to the story I tell about Bolivia: a nationalism whose racial and gendered dimensions were folded into old left projects of the mid-twentieth century, a nationalism that went on to become the substratum on which new left formations developed in the early twenty-first century. As a bedrock, however, nationalism was never as stable as it appeared, and new movements erupted along old fissures. As exemplified by Bolivian mining cooperatives, the past can resurface in the present in ways that are neither identical to nor comprehensible apart from its historical conditions of possibility.

As the metaphors in the preceding paragraph suggest, the nation that I explore in this book is about more than people and land. I conceptualize the nation in three dimensions, extending deep into the geological layers beneath the soil, and I am attentive to the matters of which nationhood is composed. In addition to imagining the nation in three dimensions, this approach is driven by an attention to the interplay between nation and nature. While critical approaches to the study of nature and nation have tended to focus on living natures—flora and fauna—the nonliving subterranean has been an important site of sociocultural and political economic production in Bolivia. Silver, tin, and natural gas have been the literal and metaphorical bedrock of the nation for the past five hundred years, with promises of lithium mining now figuring large on the horizon. Each of these materials has emerged from the subsoil already entangled with distinct nation-building projects, each shaping and shaped by a new constellation of social inclusions and exclusions, typically drawn along raced and gendered lines. As historian Kevin Young (2017) has compellingly argued, modern Bolivian nationalism has always been a form of resource nationalism. Resources, I would add, are always socially marked; they are raced and gendered before they ever see the light of day.

My focus is on tin, a commodity extracted from metalliferous veins laced throughout the mountains between the city of Oruro and the northernmost provinces of Potosí. Although the history of the tin industry is often told in strictly economic terms, its emergence was concomitant with a new vision of nationhood, most often associated with the National Revolution of 1952. This revolution was led by unions of male mestizo miners, and the nation they imagined is not just remembered; it has also left material traces across the Bolivian landscape. In the tin mines, cooperative miners labor in the same hallowed/hollowed shafts that were once excavated by unionized tin miners. They contend daily with slag heaps, rusted machinery, and abandoned company buildings. As cooperative miners extract the remnants of a twentieth-century metal, their bodies, social organizations, and political desires are shaped by both the material qualities of the ore and the values instilled within.

The material history of nature is thus integral to a story of nationalism in Bolivia that, I argue, helps to explain the tensions embedded in the new left. Subterranean materials, some sparkly and some grimy, are ultimately as constitutive of plurinational Bolivia as its citizenry and administrative processes. In this book cooperative miners are guides to the subterranean and its connections to both national political economy and national cultural politics. From the underground encounter of worker and rock, the place where a miner can push a pick into a slim black line of tin ore, spirals a set of relations and memories that fundamentally shape the meaning of nation in Bolivia today. *Subterranean Matters* follows this spiral.

I have delayed writing these acknowledgments until the last possible minute, both because I'm afraid I will miss someone and because the list of people I want to thank keeps growing. But no matter how long it gets, I will always remain most grateful to the people from Llallagua-Uncía who supported my research, particularly the cooperative miners but also many *compañeras* and *compañeros* from Radio Pío XII, the Universidad Nacional Siglo XX, and the COMIBOL Archives in Catavi. I am especially thankful to the woman I call Demetria, who was always my strongest advocate in Llallagua; to Tata Max's generous family in Uncía; and to all the miners who endured my continuous questions and looked after me underground. There would be no book without them. My profound thanks.

If I retrace my steps, I believe the kernel of an idea for this book formed in conversation with Tom Perreault in 2011 in a café in Cochabamba, where I was working on an earlier research project. Thank you, Tom, for pointing me in this direction and for later introducing me to mining on the altiplano—a world unto itself! I owe similar thanks to Karen Bakker for the push that sent me to Bolivia in the first place. I learned of Karen's sudden passing while reviewing the final proofs of this manuscript, and was reminded that I owe a great deal to her tough but consistently supportive mentorship style. Her influence remains even as she is missed.

I am extremely thankful for all the support I received at Berkeley, especially from Jake Kosek, Donald Moore, Michael Watts, and Gill Hart. Jake's boundless enthusiasm and theoretical creativity, plus Don's generosity and seemingly infinite wisdom, kept me going through the long research process. I am also grateful to Nancy Postero, who welcomed me into her group of *bolivianistas* at the University of California, San Diego, and to Nancy Lee Peluso, for inviting me to join her team of small-scale mining researchers. Thank you all for being such incomparable mentors.

In a more informal but no less influential way, I am deeply indebted to all the people at Berkeley with whom I shared seminar spaces, writing groups, street protests, and dance floors. Foremost among these are the lovingly named Geogrababes, a group that includes Camilla Hawthorne, Meredith Palmer, Erin Torkelson, and Mollie Van Gordon, without whom I would never have finished the PhD, let alone the book. Having my WhatsApp fire at all hours of the night while each of us conducted research in different time zones was a comfort throughout the sometimes-lonely fieldwork process. Along with Erin, Camilla, and Meredith, I later benefited from being in a writing group with John Elrick, Brittany Meché, Jen Rose Smith, Alex Werth, and Ashton Wesner, and in other contexts I learned tremendously from Angela Castillo, Jimena Diaz Leiva, Brian Klein, Matt Libassi, Juliet Lu, Bridget Martin, Jeff Martin, Sebastián Rubiano Galvis, Julia Sizek, Alessandro Tiberio, and Shuwei Tsai. Thank you for making the process beautiful and so worth it.

In La Paz and Cochabamba, I have been grateful for the support of other researchers and friends, including Penelope Anthias, Rocio Bustamante, Carola Campos Lora, Vladimir Díaz, Susan Ellison, Linda Farthing, Thomas Field, Kirsten Francescone, Karina Guzman, Sarah Hines, Aiko Ikemuro Amaral, Amy Kennemore, Elizabeth López, Angus McNelly, Hans Möeller, Silvia Molina, Luis Oporto Ordóñez, Pablo Poveda, Hernán Pruden, Yara Terrazas Carafa, Bill Wroblewski, Alfredo Zaconeta, Gabriel Zeballos Castellon, Adriana Zegarra, and many others. I owe the most to the amazing Yara and her family, who welcomed me into their home whenever I landed in La Paz, and to Amy and Bill, two magnificent housemates and collaborators. Institutionally, I have also benefited from the support of the research center CEDLA (Centro de Estudios para el Desarrollo Laboral y Agrario; Research Center for Labor and Agrarian Development) and the COMIBOL Archives in El Alto. Thanks to everyone who so generously shared their time, knowledge, and the occasional Huari with me.

I have been lucky to find brilliant and encouraging colleagues at Rutgers, in the Geography Department and beyond. For intellectual inspiration and/or general camaraderie, shout-outs are owed to Karishma Desai, Priscilla Ferreira, Nate Gabriel, Asher Ghertner, Zeynep Gursel, David Hughes, Jenny Isaacs, Mazen Labban, Robin Leichenko, Preetha Mani, Melanie McDermitt, Paul O'Keefe, Will Payne, Åsa Rennermalm, Kevon Rhiney, Dave Robinson, Kevin St. Martin, Zakia Salime, Laura Schneider, Genese Sodikoff, Mary Whelan, Willie Wright, and Omaris Zamora. Many thanks to Michael Siegel for the excellent maps that appear in this book, and to Cleo Bartos and Kelly Bernstein for keeping our departmental ship on course. I have also learned enormously in conversation with a fabulous group of graduate students. Although there are too many to name, I particularly want to acknowledge my first three doctoral advisees: Liana Katz, Jamie Gagliano, and Daniela Mosquera-Camacho. There are few people who can balance intellectual and political commitments with their aplomb, and I am honored to witness their creative processes.

I wrote a large chunk of this book (including this sentence!) while in a Zoom writing room with Karishma Desai, the only person willing to wake up and tap away with me at five in the morning during the worst days of the pandemic. I later sent my often-half-baked thoughts to Jen Rose Smith, whose incisive comments were critical throughout the writing phase. She also participated, along with Penelope Anthias, Bruce Braun, and Susan Ellison, in the book workshop that resulted in the eventual transformation of a collection of chapter drafts into a full manuscript. My heartfelt thanks to all four participants.

In reverse-chronological order, funding for this book came from an American Council of Learned Societies Faculty Fellowship, a Mellon Foundation / American Council of Learned Societies Dissertation Completion Fellowship, a Pierre Elliott Trudeau Scholarship, a Social Science Research Council Dissertation Proposal Development Fellowship, and a Social Sciences and Humanities Research Council of Canada Scholarship. I am very thankful for all these fellowships, and I am especially grateful that the Trudeau Foundation events created frequent opportunities for me see my close friend and geography comrade Cynthia Morinville. Thank you for all the conversations in so many cities across North America, Cynth.

My parents have been an unwavering source of encouragement for as long as I can remember. They kept their cool when I announced I was switching from an undergraduate science major to a suspicious interdisciplinary studies major, when I decided to pursue a PhD in a discipline that doesn't even exist in many universities, and when I told them I would be spending large stretches of time in underground mines with minimal safety precautions. I have always appreciated that they celebrated my unusual and certainly unlucrative life decisions, sometimes against their better judgment. I am similarly grateful for my sister, Elizabeth, whose capacious intellect has always inspired me even as it left me in the dust, and to my aunt Monica, who connected me with my own family's Cornish tin-mining roots. Thank you all for the many leaps of faith you have had to take with me.

Finally, I am so grateful for my partner, Jesse Rodenbiker, and our daughter, Akira. For much of the time I was living in Bolivia, Jesse and Akira were on the exact opposite corner of the planet, in southwestern China, with a twelve-hour time difference. Every time I went underground—which would happen at around 8 p.m. China time—Jesse would stay awake until he received word that I had made it out again. Precisely because of its impracticality (after all, what could he have done if he hadn't heard from me?), I can think of no gesture more loving. Thank you for being so wonderful. And Akira, the daughter I never knew I'd have and whom I couldn't imagine living without, thank you for putting up with my surprise lectures and long travels, and for keeping me on my toes with your earnest curiosity about the world. Wherever we go, you are both at the center of my heart.

The news hit like an electric shock, tearing through regularly scheduled programming across the media spectrum. In La Paz, a city accustomed to arranging its daily movements around parades, marches, and road-blocks, it is hard to generate a scandal worthy of more than a groan. But this was different. In August 2016 groups of small-scale miners known as *cooperativas mineras* (mining cooperatives) had erected a roadblock in Panduro, a town along a well-traveled highway between the cities of La Paz and Oruro. Several days into the roadblock, the situation had become violent. Everyone was glued to TV screens as grisly photos and videos began to surface on Twitter and Facebook, the images grainy and the voices barely audible.

The deputy minister of the interior, Rodolfo Illanes, was dead. Murdered. He had gone to the roadblock to negotiate its end, and his body was discovered in the wee hours of the morning of August 26, wrapped in a sheet and dumped on the side of the road. The cameras zoomed in on his face, bloody and swollen. A video filmed before his death showed him talking on a cell phone at the center of a dense ring of angry miners. He appeared to be begging the person on the other end of the line for help. Near the videographer, someone yelled, "A ver un palo, yo le voy a hacer gritar" (Get me a stick, I'm going to make him scream).[1] Later, the coroner announced that Illanes had indeed been tortured for six to seven hours before he was beaten to death, with the final blow delivered by a rock to the back of the head. Rumor had it that part of the torture had involved exploding dynamite near his ears. The picture of Illanes's car that appeared on Twitter revealed that it had similarly been burned with explosives, the tires completely melted off and the hood curled upward from the heat. Dynamite, the signature tool of the miners, was also their signature weapon.

The events that had provoked this protest were complicated. Most news outlets, however, focused on just one facet of the story: cooperative miners were angry about a proposed modification to the Ley General de Cooperativas (General Law of Cooperatives; Law 356, originally passed in 2013), which was slated to recognize the right to unionization for workers employed by all "cooperative societies." Mining cooperatives had a reputation for compensating their workers poorly, and journalists concluded that the miners were trying to defend their system of exploitation against the threat of unionization.

But this explanation raised more questions. Legally categorized as "productive cooperatives," the miners had never been allowed to hire third-party workers (except for administrative and technical support); it was the service cooperatives, which mostly provided water and phone connections, whose hiring practices were under scrutiny with the revised law. Despite not being targeted, however, cooperative miners had adopted the struggle as their own. They used it as a platform to release a ten-point *pliego* (list of demands), only one of which was related to the General Law of Cooperatives. The other demands included relaxed environmental standards, the extension of electrical lines to all mining cooperatives, and a modification to the Ley de Minería y Metalurgia (Law of Mining and Metallurgy; Law 535) that would allow mining cooperatives to partner with transnational mining corporations.[2]

Only with the release of this *pliego* did most Bolivians get an inkling of the depth of discontent that was brewing between mining cooperatives and the state under the leadership of leftist President Evo Morales, the country's first Indigenous-identifying president. By most accounts, Evo was wildly popular, and cooperative miners had long been among his most loyal constituents. What were supporters of Evo's celebrated "process of change" (*proceso de cambio*) doing murdering a member of his cabinet? Why did so many small-scale miners care about Evo's process of change to begin with?

GOING UNDERGROUND IN PLURINATIONAL BOLIVIA

The events just described took place while I was conducting research for this book, and they capture some of the political ambiguities that made the work so challenging. Although I had a lot of questions at the outset of my research, I was confident that I knew, at least, what mining cooperatives *were*. In preliminary conversations with Bolivian researchers and activists that began in 2011, mining cooperatives had been described to me as collectives of small-scale miners that were "cooperative in name only," a phrase that my interlocutors used to emphasize how *co-operativistas* (cooperative miners) individualized both the risks and the profits of mining. While this kind of individualization is common among artisanal and small-scale miners worldwide—for instance, among the *galamseys* of Ghana and the *garimpeiros* of Brazil—the word *cooperative* typically suggests some kind of economic redistribution that, my interviewees assured me, could not be found within mining cooperatives.[3] From these conversations, I also learned that mining cooperatives at the time had a total estimated membership of around 122,000 nationwide, a figure that would likely quadruple if one included ancillary employees and dependent family members (Mamani 2018).[4] Finally, I understood the basic contours of their historical emergence. The number of cooperative miners had exploded in the 1980s and grown steadily thereafter, a pattern that aligned with the global spread of small-scale mining in the wake of both economic liberalization and technological changes that decreased demand for "unskilled" mining labor.[5] In Bolivia specifically, the rise of mining cooperatives was typically dated to 1985, when a suite of neoliberal policies resulted in the closure of the country's nationalized tin mines. Suddenly bereft of employers, cooperative miners seemed to embody the neoliberal ethos: they plundered their own mountains and gambled their own lives in the hopes of striking it lucky.

But the longer I worked with cooperative miners, the harder it became for me to clearly locate them in the political, economic, and historical landscape described in these early interviews. In fact, I was eventually convinced that *the* defining feature of cooperative miners is that they are difficult to categorize, at least in the categories typically used to understand Latin American politics. Many *cooperativistas* are descendants of unionized miners, but they do not ascribe to union organizing or the traditional left. Although many of them maintain strong ties with Indigenous communities, only some identify as Indigenous. They often spend part of their year doing agricultural work on family land and sometimes participate in campesino (peasant) unions, but they are far from subsistence farmers since mining necessitates participation in the market. They are sometimes classified as petit bourgeois entrepreneurs, but in practice most of them treat their underground mining sites as family plots rather than capital investments. Although they are often characterized as the personification of "savage neoliberalism," they align themselves with the anti-neoliberal social movements that culminated in Evo's election in 2005. At a general level, since they overlap with and influence collectives that do not have any official ties to small-scale mining, it seemed increasingly untenable to me that they should be bracketed apart from the larger fields of Bolivian political and economic organizations. It is hard to understand mining cooperatives without understanding these other institutions—and vice versa.

From my outsider's perspective, cooperative miners' support for Evo seemed particularly counterintuitive. Evo, an Aymara speaker with a global reputation for environmental advocacy who enacted laws to protect Pachamama (Earth Mother) and who made radical speeches about climate change while wearing patterned knit sweaters, is far from an obvious presidential choice for a group of people economically dependent on nonrenewable resource extraction. Yet *cooperativistas* supported his administration for most of his time in office (January 2006–November 2019)—a fact that has been obscured within most existing accounts of Evo's presidency, which tend to emphasize the involvement of Indigenous and campesino federations, trade unionists, and urban informal (popular) workers. While they were never exactly the face of Evo's electorate, mining cooperatives nevertheless made up its raggedy extremities, the places where the rallying call for *el pueblo* (the people) was stretched the thinnest. Given the imperfect fit between cooperative miners' aspirations and Evo's political program, the conflict that resulted in Illanes's death becomes slightly clearer. The question of why cooperative miners supported Evo to begin with, however, remains murky.

In this book I contend that cooperative miners are emblematic of the tensions that characterized Evo's Bolivia and the Latin American left more generally. Evo's administration was the centerpiece of Latin America's "new left," a rising tide of left-leaning governments that ascended democratically to power across Latin America in the early 2000s. Evo epitomized the apparent novelty of the new left, as he appeared to embody a synthesis of working-class and Indigenous politics. This synthesis, however, was never free of contradiction, neither in Bolivia nor in other politically aligned countries (which included, in the late 2000s, Argentina, Brazil, Chile, Ecuador, Paraguay, Uruguay, and Venezuela). These contradictions often manifested within the mining cooperatives themselves, which were fickle allies of workers' unions, Indigenous federations, and urban guilds (*gremiales*). For this reason, I wager that the paths traced by cooperative miners can offer fresh insights into the constitution not only of the Bolivian left but also of the Latin American left more generally.

With cooperative miners as guides, *Subterranean Matters* argues that the tensions between the old left and the new left can be more adequately understood as a tension between competing senses of nationalism, all of which are entangled with the meanings and matters of the subterranean.[6] The word *patria* is important for understanding this entanglement. *Patria*, which can be literally translated as "fatherland," is used more frequently in Bolivia than its synonyms *país* (country) and *nación* (nation). Derived from the Latin word *pater* (father), *patria* suggests a claim of belonging, or congruence between national citizen and national land, which is inherited through patrilineal succession. A gendered vision of biological descent is thus equated to a territorialized collective identity. *Pater* is also the root word of *patrimonio* (patrimony), which in Bolivia is frequently used to discuss natural resources that supposedly belong to all citizens, constitutionally and discursively. If *patria* is the claim to territorial belonging, *patrimonio* is the territory's inherited wealth. The etymological entanglement of *patria* and *patrimonio* speaks to the conceptual entanglement of nation and nature, threaded together by traditions of masculine and racial inheritance as much as political machinations. The question of who belongs to *patria* is inextricable from the question of who has a right to make decisions about or benefit from its natural resources.

The natural resources that matter most in Bolivia, and in this book, are buried deep underground. Indeed, the subterranean is the main protagonist of this book, despite the many cooperative miners that fill its pages. In its simplest form, my central contention is that the meanings and matters of the subterranean are fundamentally constitutive of the nation and senses of nationalism in Bolivia. Although most theories of nationhood play out along the horizontal plane of land—often conjuring a primordial, rooted connection between people and soil—I insist that the rocky, infertile depths of the earth have subtended this national imaginary in multiple ways. The literal bedrock of the nation has been variously constructed as the sovereign realm of the state, as a shared inheritance, and as a collection of natural resources awaiting extraction. In these forms the subterranean has permitted the nation's political, economic, and sociocultural unity, even while remaining invisible to most Bolivians.

Questions of nation and subterranean nature were central topics of debate throughout Evo's time in office. When he first came to power, his administration seemed to reject not only neoliberal economics but also many aspects of liberal political theory, including assumptions about the unity of the nation. In 2009 Evo ushered in a groundbreaking new constitution (the CPE: Constitución Política del Estado; Political Constitution of the State) that transformed the Republic of Bolivia into the Plurinational State of Bolivia. This change signaled a fundamentally reorganized relationship between the state and the many Indigenous, intercultural, and Afro-Bolivian communities whose traditional territories are crossed by Westphalian lines (CPE 2009, especially Articles 3 and 5; García Linera 2014).[7] A reorganized vision of nation was accompanied by a new approach to nature: Indigenous values, such as harmony and reciprocity, featured prominently in the constitution, and Indigenous conceptions of Pachamama formed a new centerpiece of environmental law. In international media outlets and many academic discussions, the consensus was that Evo's administration was enacting an Indigenous environmentalism that would be salutary not only for the country but also for the planet. While a few decades before anthropologist Orin Starn (1991) could critique Andean scholars for "missing the revolution" in Peru because they were too focused on cultural stasis, in the early 2000s Andeanists began flocking to Bolivia precisely to witness Indigenous revolutionary struggle.

As with most of the other new left administrations, however, Evo's ability to implement the social programming that Bolivia so desperately needed—universal health care, rural electrification, and social grants, among others—depended on resource rents. Evo's rise to power coincided with a commodity boom that was already triggering rapid growth in the country's mining and natural gas sectors. One of his first acts on being sworn in was to nationalize the natural gas sector. Although the extent to which this was a "true" nationalization can be debated, it enabled Evo's administration to harvest enough gas rents to lift millions of Bolivians out of poverty.[8] Between 2005 and 2012, Bolivia's extreme poverty rate dropped from 38.2 percent to 21.6 percent, a success that was acknowledged by even the world's most neoliberal institutions (International Monetary Fund 2014, 77).[9] Bolivia was not alone in this strategy: to greater and lesser extents, resource rents were used to finance pro-poor programs in Ecuador, Venezuela, Brazil, and Chile, among others. As Uruguayan sociologist Eduardo Gudynas (2009) influentially described it, the new Latin American left developed an economy based on "neo-extractivism," understood as state-led extraction used to benefit "the people" in ways that legitimated both the administrations and the extractive economy itself.[10]

But the steady expansion of extractive sectors, including not only natural gas but also minerals and industrial agricultural products like soybeans, raised the ire of more than a few of Evo's initial supporters, especially Indigenous federations and environmental advocates. In 2011 a massive conflict emerged after Evo announced plans to build a highway from the tropics of Cochabamba to the Brazilian border, slated to pass directly through lowland Indigenous territories and a national park called TIPNIS (Territorio Indígena y Parque Nacional Isiboro Sécure; Isiboro Secure Indigenous Territory and National Park).[11] After that, Bolivia's two major Indigenous federations—CONAMAQ from the highlands (Consejo Nacional de Ayllus y Markas del Qullasuyu; National Council of Ayllus and Markas of Qullasuyu) and CIDOB from the lowlands (Confederación de Pueblos Indígenas de Bolivia; Confederation of Indigenous Peoples of Bolivia)—publicly denounced the gap between Evo's discourse and action. The TIPNIS conflict was not an isolated incident, although it was an important flash point. Business had been booming for hydrocarbons, minerals, and agricultural products, encouraged by high prices and a supportive government. Indeed, the president who was supposed to represent the indigenization of the state became, in his later years in office, ironically popular among agribusiness owners in lowland Santa Cruz.[12]

These environmental conflicts presaged Evo's eventual downfall. In October 2019 Evo's fourth consecutive presidential victory was challenged by multiple groups citing voting irregularities and constitutional term limits. Protests turned violent, and leaders from across the political spectrum—including workers' unions, Indigenous federations, far-right coalitions, and the army—demanded that Evo resign. He complied, but so did the next four people in line for power. In the political vacuum that emerged, the little-known politician Jeanine Áñez, second vice president of the Senate and fifth in line for the presidency, assumed the presidency. Áñez belonged to an obscure right-wing party from the lowland department of Beni, and one of her first actions on gaining power was to "return" an enormous Bible to the presidential palace. With the support of Bolivia's traditional elites, particularly in the lowlands, Áñez stayed in office for just shy of one year (November 12, 2019–November 8, 2020). Although the elections of 2020 installed a former member of Evo's cabinet, Luis ("Lucho") Arce, an economist who promised to restart the "process of change" that Evo had set in motion, the Áñez administration had already dealt a serious blow to the Bolivian left.

The reactionary forces that seized power in 2019 did not manifest out of thin air. Instead, they articulated a particular vision of nation and nationalism that was—although by all accounts dormant during the Evo administration—always active just below the surface. I mean this metaphorically but also quite literally. Competing senses of nationalism in Bolivia, I argue, are spatialized along the vertical axis. Indeed, as I elaborate in chapter 1, the contemporary tension between Indigenous plurinationalism and nationalist extractivism is the most recent iteration of a much older tension between the politics of land and the politics of the subsoil in Bolivia. Since the colonial era, land has existed in the plural, meaning that it could be owned privately or collectively and was frequently imbued with place-specific meaning. Bolivia's subterranean depths, however, were produced as a national space and infused with meanings stemming from nineteenth- and twentieth-century nation-building projects. Regardless of the chaos on the surface, the subterranean was the bedrock on which successive iterations of nationhood established their legitimacy. Each wave of nation building has invested new hopes in the subterranean, but previous dreams remain sedimented below.

The subterranean has thus become an archive not only of geological histories but also of nationalist aspirations. Cooperative miners are boundary crossers, and not only in terms of the challenges associated with categorizing their economic or ethnic-racial identities: cooperative miners physically cross the boundary between soil and subsoil on a daily basis. In so doing, they expose the raw friction between these realms as much as their unexpected unity. Perhaps because of these transgressions, mining cooperatives were already a sore spot in the national imaginary long before they murdered Illanes in 2016. Going underground with cooperative miners is a way of seeing (and feeling and smelling) how earlier national imaginaries continue to impinge on the present.

SUBTERRANEAN MATTERS AND VERTICAL SPACES

Throughout this book I develop my analysis of mining cooperatives, nationalism, and the subterranean in conversation with Bolivian theorists. By "Bolivian theorists," I mean both Bolivian scholars, or those who have recorded their interpretations in written form, and the numerous organic intellectuals who shared their interpretations with me over coffees and beers, in buses and taxis, at home and at work. Even when I did not fully agree with their analyses, I tried to build my arguments in conversation with theirs.

More often than might be expected in the US academy, this meant being in conversation with revolutionary lefts of various stripes: Marxism, Trotskyism, Gramscianism, and *lorismo*.[13] Latin America has a rich tradition of radical political economy, and this seems particularly true in Bolivia, where Marxian historical materialism has provided a powerful explanatory framework for understanding the long relationship between colonial extraction and imperial capitalism. This influence is not confined to universities, as Marxism and Trotskyism inform the framework used by regular citizens to evaluate the benefits and drawbacks of various economic programs. Studying mining in Bolivia means having a firm grasp on historical materialism; otherwise, interviews quickly devolve into remedial lessons for theoretically impoverished gringos. This happened to me frequently, even though I had already received (what I thought was) rigorous training in dialectical and historical materialisms. I can only imagine that it would have been harder to gain my interlocutors' approval without this prior experience.

At the most fundamental level, historical materialism describes a method of political-historical analysis that takes society's "material relations" as the agential motor of historical change. What counts as material relations, however, varies significantly across the literature. Basic invocations of historical materialism describe a dialectical relationship between workers and capitalists, in which capitalists own the means of production (resources, tools, land) and workers own only their own labor. Although they meet on structurally uneven footing, both capitalists and workers need one another: without workers, capitalists could not produce; without capitalists, workers would be unwaged. As workers come to understand both their collective plight and the power of collectively withheld labor, however, tensions mount that can only be resolved through revolutionary struggle.[14]

In this simple rendering, workers' anticapitalist struggles can sometimes appear to occur naturally because of the structural conditions of mutual dependence in which capitalists and workers coexist. This version of historical materialism is frequently critiqued for its blinkered focus on economic relations (to the exclusion of colonial, raced, gendered, sexual, and other forms of oppression) and for its adherence to a progressive sense of improvement over time, as inherited from European Enlightenment philosophies. But to describe historical materialism in these terms alone does not do justice to the theory or its methodological entailments. A more nuanced approach—or one more faithful to Marxist philosophy—also explores how individuals and groups of people are themselves historical products, shaped in sensuous interaction with the world around them.[15] For Karl Marx, this formation happened, above all, through the labor process. As workers transformed natural resources into commodities, they were in turn transformed. Their bodies changed to accommodate the labor process, but even more than that, the labor process shaped their interpretations of the world and of their position within it. What emerged through this process, wrote Marx, was consciousness.[16] With this added layer, the materialism of historical materialism refers both to dialectical relations between workers and capitalists (i.e., between social forces) and to dialectical relations between workers and nature (socionatural forces).

Like historical materialist accounts, this book focuses on labor as a key site of encounter between human and nonhuman nature. I differ, however, in how I historicize both human and nonhuman natures. While the material conditions of labor have always been important for historical materialists, their understanding of materiality does not usually accord any historicity or productivity to nature in its own right. Although the worker might be a historical subject, the materials with which and in which workers labor have typically received short shrift. By contrast, I emphasize the historical excesses that established the conditions of possibility for such encounters. I call this a *material history* rather than historical materialism, since it stresses the material-discursive and place-specific history of the materials prior to their entry into political economic relations (Marston 2021). Material history involves starting with the places of encounter between apparently discrete objects—the matters of flesh, minerals, mines, land, waste, machines, and so on—and working backward to understand how they were constituted as such. In this book I show how bodies and rocks were mattered differently by colonial, geological, and technoscientific processes long before encountering one another in the workplace. Rocks and flesh have histories that exceed economics, and they bring these histories into their economic relations, which is necessarily transformational.

In writing this, I am drawing inspiration, if not whole conceptual frameworks, from a group of literatures frequently collapsed as "new materialities." New materialists conceive of matter not as a condensed site of social relations, as with Marx's commodity, but as politically important in its own right and sometimes endowed with agential capabilities. This approach to matter disturbs Christian-Enlightenment theories of "the human" as defined by a movement out of (inert, material) nature and into (active, immaterial) consciousness, since matter itself is conceived as capable of effecting change (Bakker and Bridge 2006, 2021; Braun and Whatmore 2010). Instead of a dialectical theory of change, in which material social relations and an emerging consciousness of those relations resolve into a new (and ideally improved) set of social relations, new materialists often gravitate to theories of contingent entanglement. In this light, matter participates in the formation of emergent socionatural assemblages whose pasts and futures exceed any particular—or even any particular collection of—dialectical relations.

While this notion of historical and material excess is important to my analysis, one must tread carefully here. In privileging the material world over the immaterial, there is an inherent danger of slipping into positivist empiricism, which would involve assuming that what can be sensorially experienced is the same as what exists. As feminist science studies scholars have long emphasized, such sensorial experiences are necessarily subjective, as one's sense of the world is inseparable from one's position within it (Haraway 1988; Harding 1992). Even more, the ability to have and describe a sensorial experience (to represent it) is so intrinsically tied to humanist ontologies that it serves as an ironic tool in the posthumanist arsenal. Further, new materialist approaches run the risk of claiming an ontological "discovery" that, as Native American and Indigenous studies scholars have pointed out, is neither particularly new nor particularly comprehensive (TallBear 2017; Todd 2016). At their best, materialist approaches should be *more* politically attuned than their phenomenological counterparts, yet new materialists sometimes revel in European philosophy while ignoring that Indigenous metaphysics has long attended to relations that involve not only humans and nonhumans but also a variety of immaterial beings.

This does not mean, however, that new materialist approaches are always or necessarily politically limited. As Indigenous, Black, queer, feminist, and postcolonial scholars of nature and technoscience have shown, it *is* possible to avoid the twin traps of empiricism and smuggled humanism while examining the political constitution of race, gender, and sexuality through the animacies of everyday objects.[17] Instead of taking matter as the starting point of analysis, they historicize its emergence; instead of showing how matter shapes social relations, they explore complex entanglements in which matter and meaning, as much as subject and object, are continuously unfolding. These are the new materialities from which I draw inspiration. If historical materialism traces *historical relations between humans*, and new materialities focus on *nonhuman things in the present*, what I am calling *material history* involves a place-based historicization of how the thing came to be thingified through both human and more-than-human relations. Nature is reincorporated into history, but it is also itself historicized.[18]

This last point is key to resisting the slide toward positivist empiricism. Matter is not a natural, preexisting surface overlain by social interpretations or cultural constructs; instead, matter can be understood as an effect of power, brought into being through the very categories that regulate it.[19] Neither tin nor tin miners, for example, preexist their encounters with one another. They come into being through a variety of knowledges and practices, none of which are *only* about tin or mining. Rather, the practices and knowledges of mining emerged in tandem with ideas about nature and nation. Supposedly apolitical production processes are always already suffused with nationalist ideals and exclusions. My focus is thus on how the material stuff of nature—particularly nature that is commodified as resource and labor—shapes and is shaped by not only economic processes but also those that temporally precede and geographically exceed economic relations. In the words of Bolivian scholar Fernando Molina (2011, 12), a resource "is considered by Bolivians to be more than a simple reality of determined physical characteristics, an input or primary material. In Bolivia it signifies 'something more'"—and this "something more" matters politically and economically. The meanings that are folded into political economic processes matter because they shape the distribution of risks and benefits. The locally specific ways that economic processes produce injustices cannot be explained by recourse to economics alone. Therefore, mine is not an attempt to create a perfect synthesis of historical materialism and new materialities so much as an effort to rethink the limits and silences of the former by using a selection of tools derived from the latter.

I am particularly interested in how raced and gendered differences—apparently social differences—are constituted through matters that are not only part of "nature" but also detrimental to human life: silica dust, noxious gases, refining chemicals. Throughout this book, but especially in chapters 3 and 4, I explore how human bodies and nonhuman natures are differentiated and hierarchically ordered in relationship to one another, particularly (but not exclusively) in the labor process. This process of differentiation is fundamentally geographic: the production of uneven space is the production of uneven socionatural formations, and vice versa. In the stories I tell, this kind of differentiation plays out above- and belowground. The mine is teeming with a variety of social life: *perforistas*, who encounter the rock face with drills; *barreteros*, who use pickaxes to pry open new veins; geologists and their scouts, who mark the passage of veins with painted arrows; and mining engineers, who devise ways to prop up the rock that looms above them. This sociality is unevenly distributed across three-dimensional space, extending deep into the earth's crust, upward into slag heaps and concentration plants, and outward across downstream fields and urban areas. Forces both natural and social have stratified vertical and horizontal space; this space, in turn, stratifies the social collectives that emerge within it.

To understand these processes, I turn to recent studies of vertical and especially subterranean spaces. Since the discipline's Marxian turn, geographers have demonstrated how capitalism is necessarily uneven, as capital produces endless spatial differentiations, between (for example) town and country, Global North and Global South, industrial neighborhood and residential neighborhood, farmland and wilderness.[20] This line of inquiry has historically focused on processes taking place across the surface of the earth. Recently, however, a burgeoning literature is challenging geography's disciplinary "land bias" (Peters, Steinberg, and Stratford 2018, 2), including within geographic deployments of Marxian theory. Collectively described as constituting a "vertical" or "volumetric" turn, studies produced in this vein have gone in several different (literal) directions. Some scholars have focused on atmospheric space, examining high-rise construction, drone warfare, urban air quality, the politics of wind power, and even extraterrestrial mineral speculation.[21] Others have gone underwater by exploring the spatial dimensions of fisheries, deep-sea mining, and the construction of artificial islands.[22] But the vertical space that most interests me is that which extends beneath our feet. As a growing number of scholars have shown, the subterranean cannot be understood as a space apart; what happens belowground shapes and is shaped by socioenvironmental processes that play out across the surface of the earth. The production of subterranean space is a constitutive part of globally uneven development.

As a three-dimensional space, the subterranean has become a shared site of interest within several apparently distinct conversations. First, political ecologists and environmental justice scholars are increasingly exploring how control over the subterranean figures in the uneven distribution of environmental resources and environmental hazards.[23] Second, political geographers and some heterodox political scientists have observed how strategies to "secure the subterranean" play out in struggles for territorial sovereignty, whether this involves surveilling for cross-border tunnels, monitoring volcanoes that might threaten existing borders, or producing detailed maps of geological formations found within national territories.[24] Third, cross-disciplinary interest in the Anthropocene has encouraged scholars to think about how the subterranean is at once an archive of deep planetary time and a repository of resources whose combustion has not only altered global climatic patterns but also shown up in the geological record as the (hotly contested) mark of a new geological era.[25] Finally, pushing back against the large-scale and sometimes-totalizing tendencies of the first three conversations, a fourth group of scholars is showing how the subterranean is always more than a repository of resources, an extension of sovereign power, or an archive of planetary change.[26] Instead, this final group shows how the subterranean is deeply meaningful within and beyond capitalist machinations and has long been integrated into human social worlds as sacred and recreational sites, life-giving aquifers, and domestic spaces, to give just a few examples.

I work across all the above conversations, but I focus on the second and fourth. I am interested in how the Bolivian subterranean was produced as national (state-owned) territory through histories of global colonialism, capitalism, and imperialism, but I am also interested in local experiences of laboring and living in this supposedly (but never fully) national space. The rocks and the people who meet at the site of labor are both the products of multiple histories, and these histories shape the uneven patterns that continue to unfold across three-dimensional space.

STRATA OF NATURE AND NATION

My goal in this book is to rethink contemporary senses of nationalism in relation to the matters and meanings of the subterranean. Such a goal implies rethinking historical origins: a material history of Bolivia must begin prior to the nation itself. While post- and decolonial scholars have long explored connections between colonialism and nationalism in Latin America, the histories charted in this section focus on the role of nature—particularly the subterranean—in mediating this relationship in Bolivia.

In the early sixteenth century, present-day Bolivia was part of the Incan Empire, and the basic unit of government was the ayllu. Ayllus are territorialized communities with nested, rotational political systems and noncontiguous lands that anthropologist John Murra (1972) famously described as "vertical archipelagos."[27] Although they predate the Incan Empire, by the sixteenth century Andean ayllus had mostly been incorporated into a network of Incan tributaries, and their members were expected to provide labor to the Incan Empire for a set number of days per year. This obligation was called the *mit'a*—which means "turn" in Quechua, the lingua franca of the Incans—and those performing it were called *mitayos*. Among other obligations, *mitayos* labored in silver mines, the largest of which was Porco, a mine that remains operational today (Galeano [1973] 1997, 21).

When Spanish conquistadores climbed up from the Pacific coast into the Andes in the 1530s, they were looking for gold, but it did not take them long to settle for a slightly less precious metal. In the oft-recounted origin story of colonial silver mining, Quechua herder Diego Huallpa was warming himself by a fire on Sumaj Orcko (Beautiful Hill) of Potosí in 1544 when he noticed that the rocks under the embers were glittering with molten silver. It seems likely that most Quechua residents of the area were already aware of the mountain's riches—it was, after all, less than forty miles from the Porco mine—but, according to the story, it was Huallpa who brought the deposit to the Spaniards' attention (Bakewell 1984). Once aware of the wealth the mountain contained, the Spanish moved swiftly to extract and export silver from the place they renamed the Cerro Rico (Rich Hill).

At first, the Spanish relied heavily on Incan technologies and governance structures, adopting an indirect form of rule that focused on extracting tribute rather than reorganizing land and labor.[28] This colonial governance system changed when the yields from the Cerro Rico began to fall. The surface deposits had already been ransacked, and the mountain's internal ores were not as easy to access or to smelt. But in 1554 the Spanish merchant Bartolomé de Medina developed the "patio process," which used mercury amalgamation to separate silver metal from ore, and in 1563 a huge source of mercury was discovered in Huancavelica, which is located in present-day Peru. Responding to the labor needs of this new technique, in 1569 Viceroy Francisco Álvarez de Toledo announced his plan for *reducciones*, or resettlement programs, which forced dispersed Indigenous households across the Andes into concentrated communities where their labor could be more easily taxed. This labor was extracted under nearly the same name as in the Incan system—the *mita*—but it was used almost exclusively in the silver mines, and the conditions were appalling. At any given time, one-seventh of all adult males were expected to be working in the silver mines of Potosí; often their wives and families went with them. The inhuman working conditions resulted in the deaths of more than eight million people.[29] As a colonial territory, Bolivia was thus forged in a crucible of silver and blood, and both were rendered from Indigenous bodies.

Against a backdrop of widespread Indigenous and Afro-Latinx insurrections, Latin American countries began to claim independence from Spain in the late eighteenth and early nineteenth centuries.[30] With the notable exception of Haiti, however, Latin American–born descendants of Spanish colonizers, known as *criollos* (Creoles), claimed power in most of the new nation-states. In Alto Peru—as Bolivia was known in the colonial era—the rise of Creole nationalism threatened an arrangement that anthropologist Tristan Platt (1982) calls the "pact of reciprocity," in which Spanish colonizers had granted ayllus self-governance on collectively held lands in exchange for tribute to the Crown. In the colonial era, the division between Spanish and Indigenous peoples was spatial, juridical, and financial: there were two different legal codes and taxation systems that corresponded with separate places and peoples. When the Bolivian Creole elite won independence in 1825, they sought to overcome this "dual republics" system by dismantling Indigenous landholdings and creating rural land markets for large-scale agriculture, but they were met with widespread Indigenous resistance. The newly minted Bolivian state, moreover, was cash poor following the long independence war, and tax revenue provided by communal Indigenous landholdings was its prime source of income. The push to liberalize abated, and the dual republics model continued well into the twentieth century (Larson 2004; Rivera Cusicanqui 1987).

By this time, elite theories about race, nature, and nation were developing newly "scientific" dimensions in Latin America. While European eugenics was creeping into nation-building projects around the world, the uptake of these theories was unique to the racial reality of each country. In Latin America, Argentina adopted the most strident project of social Darwinism; Brazil aimed for whitening through European immigration and education programs; and Mexico, Peru, and Bolivia developed concerted projects of mestizo nation building.[31] *Mestizaje* translates literally as "miscegenation," and prior to the twentieth century, such racial mixing was understood in largely degenerative terms, despite proliferating lists of racial "types." By the 1920s, however, *mestizaje* was being reimagined as the optimal combination of supposedly Spanish, Indigenous, and African attributes. In his book *La raza cósmica* (The cosmic race), which defined Mexican mestizo nationalism for decades to come, José Vasconcelos argued that the racial diversity and tropical climate present in Latin America would produce "the definitive race, the synthetical race, the integral race, made up of the genius and the blood of all peoples and, for that reason, more capable of true brotherhood and of a truly universal vision" ([1925] 1979, 20). This was the era of racial eugenics in Europe and around the world, but Latin American leaders were often more receptive to Lamarckian than Darwinian eugenics, meaning that they imagined the possibility of racial "improvement" through environmental changes rather than strictly genetic inheritance (Stepan 1991).[32] In the early 1900s, these leaders developed programs that aimed to improve their nations' "racial stock" through hygiene, bodily care, and education. All these programs focused on shaping people through engineering their environments in ways that distanced them from an external nature, as imagined by Creole policy makers. This involved a literal distancing from the earth: bare feet, earthenware pots, and dirt floors were associated with indigeneity and treated as public health concerns (Orlove 1998).[33] I explore the different theories of *mestizaje* that emerged in Bolivia in chapters 1 and 3, but the point to underscore here is that the debate about the relative merits of *mestizaje* remained an elite concern throughout the first two decades of the twentieth century.

It took a war and a revolution for Bolivian theories of *mestizaje* to be fully articulated with a broad-based popular nationalism. In an analysis that remains influential today, Bolivian sociologist René Zavaleta Mercado (1986) argued that modern Bolivian nationalism did not emerge until the Chaco War between Bolivia and Paraguay (1932–35). This war was initiated by Bolivian president Daniel Salamanca, who hoped to gain control over the Gran Chaco desert, a region that the two countries had disputed for decades. While some historians have contended that this war was an attempt to distract national attention from a dire economic situation, others have underscored that the Chaco was rumored to contain oil.[34] Regardless of Salamanca's intentions, however, Bolivia lost the war disastrously. But during three years of near-constant retreat, a sentiment that Zavaleta Mercado described as *popular nationalism* emerged from the trenches. More recently, historian Kevin Young (2017) has argued that this popular nationalism was always a *resource* nationalism, since from its inception it was concerned with wresting natural resources from foreign powers. Young defines resource nationalism as the "belief that natural resources should be used to benefit the nation" (1), but several important questions are concealed behind this deceptively simple definition. Who belongs to the nation? What benefits do natural resources offer, and how should they be distributed? Who should bear any associated costs? None of these questions was answered in the 1930s, and all of them remain salient today.

Most Bolivians *did* agree that the existing extractive regime was decidedly unfair. This was especially evident in the country's most lucrative industry: tin mining. The silver industry had played an important role in Bolivia's economy after the country's independence from Spain and had surged again in the 1870s and 1880s, but by the turn of the twentieth century, it was flagging. Bolivia's economic epicenter shifted from the silver mines of south-central Potosí to the tin mines of northern Potosí and Oruro, a shift that was so powerful that it moved the country's executive and legislative center from the temperate town of Sucre (near the silver-mining city of Potosí) to the high-altitude city of La Paz (comparatively near the tin-mining city of Oruro) in 1899. The explosive growth of tin mining in Bolivia precipitated a twenty-year economic expansion led by just three "tin barons": Simón I. Patiño, Moritz (Mauricio) Hochschild, and Carlos Víctor Aramayo (Klein 2003).

Although the financial benefits of tin mining may have accrued to a very small number of people, the tin mines were politically capricious. They had also given birth to some of the nation's strongest workers' unions, nurtured both by Bolivia's long history of artisan-led anarcho-syndicalism and by newly popularized political economic theories, especially Marxism and Trotskyism.[35] In the post–Chaco War period, a plethora of new political parties emerged, and all of them sought the support of the tin miners. One of these was the MNR (Movimiento Nacional Revolucionario; National Revolutionary Movement), which was founded by mostly middle-class students but managed to attract support from factory workers, miners, middle-class professionals, and rural smallholders. Among these groups, the tin miners' unions were the most militant and are still remembered as the revolutionary vanguard.

In 1952 the MNR led the National Revolution that established Víctor Paz Estenssoro as president. This revolution resulted in three changes that collectively reformulated the relationship between nature and nation in Bolivia. First, Paz Estenssoro nationalized all the tin barons' mines and created COMIBOL (Corporación Minera de Bolivia; Mining Corporation of Bolivia) to manage the nation's new tin reserves. Second, he introduced an agrarian reform program in 1953 that focused on redistributing land from private haciendas (plantation-style farms) to landless farmers in the highlands. All these newly landed smallholders, moreover, were to become members of state-sponsored campesino unions. While this reform was eagerly accepted in the Cochabamba valleys, where haciendas had been most widespread and landless farmers had begun to organize themselves into unions prior to the revolution, it faced steady resistance in the ayllu strongholds of La Paz and Norte Potosí (Postero 2007; Rivera Cusicanqui 2004). Third, finally moving away from the dual republics model of government, Paz Estenssoro elevated everyone to the category of citizen and universalized suffrage. While this legal transformation is usually underscored in histories of Bolivia, less frequently noted is the racial recategorization that accompanied it. In addition to becoming citizens, all Bolivians were officially classified in class terms rather than race terms. Instead of being *indios* and criollos, everyone in Bolivia became racially mestizo: *indios* became mestizo campesinos, and *criollos* became mestizo professionals.

In combination, these strategies worked to create an isomorphism between the new state, often called the "State of '52," and the citizenry it claimed to represent. The state held the subsoil, the people held the land, and all were united (in theory) by a shared origin story. Nature was gendered and racialized as a collective Indigenous mother, firmly located in the past. From her, all Bolivians had inherited the right to benefit from natural resources, but these were held in trust by a paternal state. The State of '52 conjured a postcolonial national imaginary by erasing actual Indigenous communities, including the ayllus, and their alternative claims to both land and subsoil.[36]

The plurinational imaginary that blossomed during Evo's administration owes much to reactions against the assimilationist nationalism of the State of '52 (Canessa 2007). In the early 1970s, a group of relatively deterritorialized Indigenous intellectuals near the city of La Paz began calling themselves *kataristas*, taking their name from eighteenth-century Indigenous revolutionary Túpac Katari. Adding nuance to the Marxist and Trotskyist schemas that had guided the miners' unions during the National Revolution, the *kataristas* sought to understand the "double oppression" of Indigenous peasants under both colonialism and capitalism. In the mid-1980s, Oxfam partnered with THOA (Taller de Historia Oral Andina; Andean Oral History Workshop), a La Paz–based group of *katarista* anthropologists, to understand and strengthen Bolivia's ayllus.[37] As a parade of dictators pummeled the miners' unions—fighting communist leaflets with bullets and layoffs—the Indigenous movement quietly took root.

The State of '52 depended on tin, and when the Bolivian tin-mining sector began to crumble in the 1980s, so too did mid-twentieth-century forms of government. Demand for tin—replaced in many industries by aluminum and tinplate—had fallen precipitously; at the same time, COMIBOL's reserves were nearing exhaustion, since very little prospecting had taken place in the postrevolutionary decades. The Bolivian economy, buoyed for years on the income from a single export, folded in on itself. The economic collapse coincided with—or perhaps precipitated—a tumultuous return to democratic politics after two decades of dictatorships. A series of interim presidents came through during the transitional period, ending with the reinauguration of President Víctor Paz Estenssoro, the same man who had led the country immediately after the 1952 revolution.

In the growing economic crisis, Paz Estenssoro "set out to undo what his 'social revolution' had accomplished some thirty years earlier" (Perreault 2005, 271). Within weeks of being sworn in, he had initiated the New Economic Policy, which scholars usually qualify as the first wave of neoliberal restructuring in Bolivia. The vast majority of COMIBOL mines were either privatized or closed, the currency was allowed to float against the dollar, and borders were opened to direct foreign investment. Most important for this story, twenty-three thousand of the country's thirty thousand miners were laid off (Kohl and Farthing 2006). Laid-off miners left the highlands in droves, spreading their knowledge of union organizing throughout Bolivia and laying much of the groundwork for Evo's later rise to power (Gill 2000). Many of these miners later returned to the subterranean spaces left behind by the retreating state, where they established mining cooperatives in a subterranean nature officially deemed exhausted and formally abandoned. Digging through the craggy layers of history, influenced by both memories of unionized glories past and visions of Indigenous economic futures, these cooperative miners retrace the invisible seams of ore and nation.

As the miners' unions collapsed, the *kataristas'* demands went mainstream. Gonzalo Sánchez de Lozada (Goni), a mining magnate who became president in 1993, was elected alongside Vice President Víctor Hugo Cárdenas, leader of the Túpac Katari Revolutionary Liberation Movement, and together they began to introduce reforms that might be described as "neoliberal multiculturalism," in that they recognized cultural difference only insofar as it was compatible with liberal market capitalism (Hale 2002). The Goni-Cárdenas administration passed the Ley de Participación Popular (Popular Participation Law; Law 1551), which transferred 20 percent of central state revenue to municipal governments, in 1994; rewrote the constitution to describe Bolivia as "multiethnic and pluricultural" in 1995; and introduced the Ley INRA (Ley del Instituto Nacional de Reforma Agraria; National Institute of Agrarian Reform Law; Law 1715), which created a legal category for *tierras comunitarias de origen* (communal lands of origin), in 1996. These reforms, while not exactly revolutionary, encouraged a fundamental shift in Bolivia's political orientation. For the first time since before the National Revolution, the 2001 census contained a question about ethnic-racial identification, and 62 percent of Bolivians self-identified as Indigenous.[38] At a more material level, these reforms enabled Evo's entry into politics. Supported by the *cocalero* (coca growers') unions of which he was the leader, Evo was elected within his municipality, and funds from the Popular Participation Law made it possible for him to scale up his political program to the national level within a few short years (Postero 2007).

The two social mobilizations that preceded Evo's election were directly linked to struggles over nation and nature: the Cochabamba Water War in 1999 and the national Gas War in 2003. Although these are sometimes framed as anticapitalist or anti-extractivist, they are more accurately described as anti-neoliberal and anti-imperial. The Water War united residents of Cochabamba and the surrounding areas in opposition to an American private company (Bechtel) that had taken control of the local water supply, but there was no consensus on how water *should* be managed to ensure equitable access (Marston 2015). The Gas War was sparked by the announcement that Bolivia's rich natural gas deposits would be exploited by a foreign company that was going to transport the gas through Chile, a country that has raised resource nationalist ire in Bolivia since Chile appropriated all of Bolivia's coastline in the War of the Pacific (1879–83). Evo's move to nationalize natural gas extraction in the early days of his presidency was the logical response to the Gas War, which was not opposed to extraction per se but to extraction by and for the benefit of non-Bolivians (Kohl and Farthing 2012; Perreault 2020). Colonial and imperial powers had long treated Bolivia as a resource repository, available for plundering without appropriate compensation, and Bolivians were reacting against this legacy. The protests were resource nationalist, and they resulted in the election of a president who enacted resource nationalist policies.

Of course, Evo also inaugurated the plurinational era. Resource nationalism and plurinationalism coexist in vertical space: the subsoil remains national (state-owned) territory even while the land above is divided and governed in new ways. Indeed, plurinationalism is often conceived in terms of horizontality, a word that invokes not only a flat (i.e., nonhierarchical) political relationship between the central state and its many constituent nations but also a two-dimensional understanding of the nation as a series of interlocking two-dimensional shapes. For instance, Raúl Prada Alcoreza (2007)—a Bolivian scholar and erstwhile Evo supporter—traced a spatial history of Bolivia by showing how new institutional maps of the plurinational era were drawn over existing maps of the nation. Prada Alcoreza's allusions to mapmaking were largely metaphorical, but they demonstrated the extent to which nation is imagined in relationship to land, a depthless "manipulable cartography of forces" (Gustafson 2009a, 1003). Given interwoven histories of nation and nature (particularly subterranean nature), such two-dimensional interpretations of plurinationalism occlude not only an important site of national formation but also the political economic relationship between nature and the nation-state.

FAULT LINES: COOPERATIVE MINERS OF NORTE POTOSÍ

This book is based on eighteen months of fieldwork conducted in 2013, 2014, and 2016–17, with two follow-up trips in 2022. I spent slightly more than half of my time in the two tin-mining towns of Llallagua and Uncía, located on either side of the Juan del Valle Mountain in the region of northern Potosí (hereafter Norte Potosí—see maps I.1 and I.2). When I was *not* in Norte Potosí, I was traveling throughout the Bolivian highlands and valleys, where I conducted interviews with earth scientists,

MAP I.1. Map of Bolivia. Prepared by Michael Siegel.

policy makers, and activists; visited other mining sites for comparative purposes; and worked in a variety of public and private archives.[39]

Within Bolivia, Norte Potosí is known for two things: tin mines and ayllus. From the early 1900s to the 1980s, Llallagua-Uncía was home to Bolivia's largest tin mine, which in turn nurtured some of the nation's strongest miners' unions. If Bolivian tin miners formed the vanguard of the 1952 National Revolution, tin miners from Norte Potosí were the vanguard of the vanguard. At the same time, Norte Potosí is known for its numerous highly organized ayllus, which have withstood centuries of colonial, liberal, and corporatist onslaughts. In 1874, for instance, the

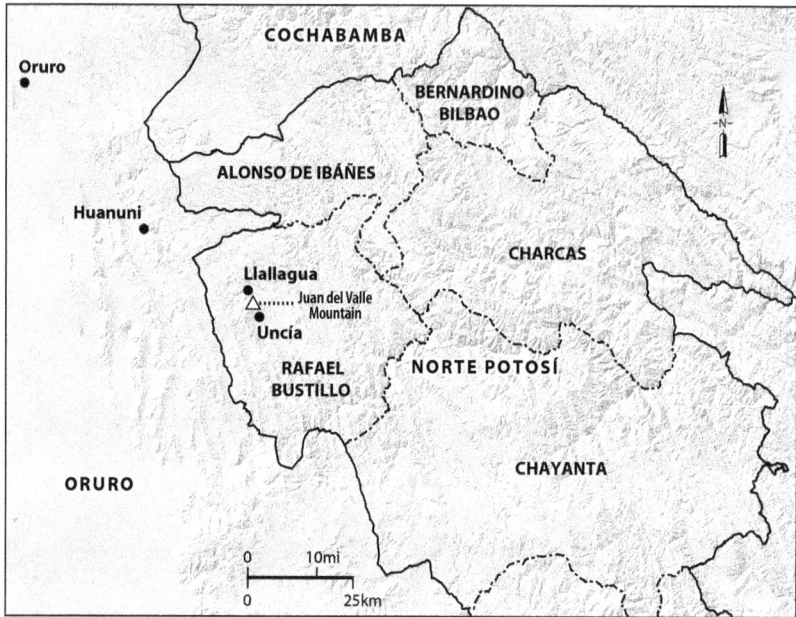

MAP I.2. Map of Norte Potosí. Prepared by Michael Siegel.

Bolivian state attempted a so-called modernization of agricultural production through ayllu dissolution, but massive Indigenous resistance in Norte Potosí prevented the program's implementation (Rivera Cusicanqui 1987); similarly, attempts to establish campesino unions after the 1953 land reform were largely unsuccessful. The image of "the warrior ayllus of Norte Potosí" is widespread and reenacted within the ayllus themselves in annual *tinkus*, which are ritualized—but genuinely violent—inter-ayllu fights in which spilled blood ensures fertility in the coming year (Le Gouill 2014).

Since the neoliberal gutting of the state mining corporation in 1985, the mine workers' unions of Llallagua and Uncía have been replaced by mining cooperatives. Today the twin towns are home to seven mining cooperatives that together incorporate some 2,600 members, all of whom belong to FERECOMINORPO (Federación Regional de Cooperativas Mineras del Norte de Potosí; Regional Federation of Mining Cooperatives of Northern Potosí). Based in Llallagua, FERECOMINORPO is one of eleven regional federations of mining cooperatives in Bolivia (one for each of the country's other eight departments and three for the mining-heavy department of Potosí). All seven mining cooperatives in Llallagua and Uncía produce tin, the metal around which the towns were initially built. In continuous creative tension with both the towns' history of unionization and the region's history of Indigenous organizing, these cooperatives continue to shape political and economic panoramas at multiple scales and multiple depths.

When I first began this project, many close friends and acquaintances in Bolivia told me that I was going to get myself into trouble. I was surprised. I had previously spent time in Bolivia studying community-run water supply systems in the peri-urban fringes of Cochabamba, and the only dangers anyone had suggested I would encounter were potentially rabid dogs. But the warnings, I found, were specific to the combination of the new topic and me as a researcher. Cooperative miners, my friends insisted, could be dangerous for women. Unionized miners of years past are remembered as masculine in a positive light: they sacrificed themselves, in both their underground labors and their armed struggle, for their families, the nation, and the global working-class. Without the direction of the unions, however, cooperative miners' masculinity is framed as wild, selfish, and indiscriminately violent. In La Paz so many people warned me about the possibility of sexual assault underground that I nearly designed a different project. Cooperative mining, they implied with their concern, necessarily produced predatory men, and these tendencies would go unchecked in the lawless space of the subterranean.

Yet aside from regular comments on my day-to-day appearance, I never encountered a cooperative miner who embodied the threat I had been warned against. A great deal of my friends' concerns for me stemmed, I believe, from the myth of white feminine vulnerability, a myth that has been used to justify immeasurable violence and is one avenue through which white supremacy is maintained. This is not to say that sexual violence or violence more generally is absent from the mines but rather that being a white foreigner shielded me from that violence rather than (as the myth would suggest) exposing me further. Indeed, the close relationship between whiteness and masculinity meant that although I spent a great deal of time with women miners, I was also able to enter men's spaces with something akin to ease (nothing ever felt fully easy). Moreover, because I was doubly foreign—both from outside the community and from outside the country—cooperative miners were less immediately suspicious of me than they were of middle- and upper-class q'aras (non-Indigenous Bolivians), whom they expected to be environmentalists or indigenistas ("indigenists," or those who romanticize Indigenous cultures or politics) come to decry the ecological and cultural contamination of mining. When I visited the Vice Ministry of Mining Cooperatives in La Paz, for instance, the representative who downloaded data onto my USB told me that he would not have given the information to a Bolivian student, since Bolivians were "not capable of understanding mining cooperatives outside of the negative discourse that circulates about them." In this case, I managed my immediate guilt by sharing the files with Bolivian researchers, but I went on struggling with my simultaneous connections with mining cooperatives and the broader community of leftist researchers.

In Norte Potosí I worked with a weekly rhythm that included regular visits to the many mine shafts scattered around the mountain, the offices of FERECOMINORPO, the offices of the seven local mining cooperatives, the local university's FPS (formación política sindical; political syndicalist formation) and mining engineering departments, the local radio station (Radio Pio XII), and the Catavi office of the COMIBOL Archives, where I pored over employee files from the 1910s to the 1930s. Once I had established myself as a regular visitor in these places, I began receiving invitations to participate in local and regional activities, such as political meetings, annual festivals, commemorative ceremonies, fairs, and parades. While participating in—and occasionally helping to organize—these events, I got to know people both connected to and critical of the cooperative mining sector. I followed up with formal interviews, usually conducted in their homes or offices, one of Llallagua's two cafés, or the tearoom of the Hotel Colonial, the unfortunately named hotel where I rented a long-term room. Inspired by Jacqueline Nassy Brown (2005), I also conducted numerous walking interviews, in which I spoke with people while they toured me through places of significance. Finally, I recorded oral histories with two cooperative miners, Demetria and Mauricio, whose candor greatly facilitated my research.[40]

I did not intend to spend as much time underground as I did, in part because these mines have a reputation for collapsing and in part because it did not seem appropriate for me, a researcher taking notes for her project, to burden a group of miners with my presence. But when I started showing up at mine shafts in the mornings in the hopes of getting to know miners while they ate breakfast and prepared for work, it became painfully apparent that everyone was far more interested in showing me their work sites than in discussing their professional trajectories. I bought myself a pair of rubber boots, a helmet, and a lamp and committed myself to going underground. In total, I spent about twenty days underground (between five and ten hours per trip). During the one-hour *pijchea* (coca chew) that always precedes a day's work, and during "crawling interviews" through the tunnels that miners jokingly refer to as an ants' nest, I got to know the work process and the workers. These underground ventures proved some of the richest parts of my research, a fact that is reflected especially in chapter 4.

I make no attempt to feign objectivity in this book. Not many foreigners come to Llallagua, and no matter how long I hung around, I remained a source of curiosity rather than a fly on the wall. In fact, some people never stopped calling me *turista* (tourist) even after I had been sightseeing for more than a year. *Gringuita, turista, choquita* (blondie): these names were used interchangeably with Andreita, a diminutive form of Andrea used as a term of endearment. The words marked a simultaneous distance and proximity that is undoubtedly reflected in my findings. I try to remind readers of this filter by situating myself clearly within the stories I tell. My account is very partial, but I hope it will be useful.

ROAD MAP

This book begins by exploring the production of *patria* and *patrimonio* in historical perspective. Focusing on three periods—early colonial (mid-1500s), early republican (after 1825), and postrevolutionary (after 1952)—chapter 1 traces the concurrent constitution of subterranean property regimes and subterranean natural resources. During these periods subterranean property law was established in conversation with forms of expertise that naturalized a particular way of seeing the subsoil and contributed to its nationalist interpretation as a shared inheritance. Most important among these forms of expertise were religious theology, in which the subterranean was envisioned as a God-given gift to the Spanish Crown, and scientific geology, in which the subterranean was envisioned as an ordered set of strata that preserved the past and yielded future wealth. I argue that the contemporary legal split between Bolivia's subsoil and surface realms can be traced back to the codification of theological and geological knowledges, which naturalized an association between the subsoil and the state while relegating divergent visions of the nation to the surface.

Starting with the second chapter, I focus primarily on the history, labors, and politics of mining cooperatives in Norte Potosí. In chapter 2

I show how the geological and chemical properties of tin informed the growth of Bolivia's tin-mining sector and, eventually, the rise of tin-mining cooperatives. I introduce the concept *material fix*, which extends David Harvey's (2001) "spatial fix" into three-dimensional space. A material fix describes successive historical attempts to rearrange labor and technology to maintain the local economy amid international price fluctuations and declining resource reserves; it also attends to the material traces left behind by past fixes. Using this concept, the chapter complicates the tale of Bolivia's 1985 neoliberalization—usually framed as the origin story of mining cooperatives—by examining how early twentieth-century and Cold War–era events created the conditions under which seven remarkably different mining cooperatives could emerge.

Chapter 3 continues this thread by examining increasing traffic between mining cooperatives in Llallagua-Uncía and the ayllus of Norte Potosí. I contend that the emergence of *agro-mineros* (agricultural miners) in the post-1985 period was important not only because it marked a moment of a regional economic diversification but also because it constituted a local indigenization of the subterranean. This chapter begins by tracing the historical separation of Indigenous campesinos from mestizo miners in Norte Potosí in relation to the two subterranean substances with which they were expected to labor: potatoes and tin ore. This history shows how livelihood practices were always already racialized, such that the recent movement of ayllu members into the mines could signify a racial transgression as much as an economic articulation. The chapter concludes by reflecting on the relationship between mining cooperatives and the Plurinational State, which has both shaped and been shaped by the rise of *agro-minería* (agricultural mining).

Chapter 4 tackles the question of individual subject formation. Irreverently borrowing from Marx's theory of consciousness, the chapter suggests that the site of labor is not only a crucible of subject formation but also ground zero for hierarchically ordering people and rocks along related axes of value. Drawing on ethnographic work conducted underground, it argues that tin's mineralogical variation—both that which occurs "naturally" and that which has been produced by a century's worth of extraction—crystallizes social stratifications among miners. Put differently, raced and gendered hierarchies are constituted in relation to the material specificities of tin as an element and as a geological formation. Minerals and miners are relationally valued in ways that shift not only spatially but also temporally, a point that is emphasized by using the concepts of *formation* and *degradation* to explore the connective tissues between geological and fleshy matters.

Chapter 5 takes as its object of analysis what I call *industrial ruins*, a category that includes old buildings, machinery, and waste rock left behind by the industrial mining practices of the twentieth century. Drawing on descriptions of walking interviews with cooperative miners and other town residents, the chapter explores how people live with and interpret these industrial ruins. I argue that although residents relate to the ruins differently depending on their own positions within the towns, their stories share a tendency to treat the ruins as monuments to the promise of temporal progress. As a result, industrial ruins—the apparently wasted remains of a previous era—continue to inspire faith in mining as key to individual and regional economic growth. Materialized in the hulls of metallic giants, mountainous slag heaps, and the sagging internal architecture of the mine itself, industrial ruins impinge on local imaginaries of the future, motivating miners underground and inflecting the politics of everyday life.

Chapter 6 returns to the national political arena—and to the murder of Rodolfo Illanes with which I opened this chapter—to explore how subterranean matters influence contemporary political dynamics, a process that is traced through two arguments. First, an abstracted sense of the subterranean as national inheritance (patrimony) undergirds dynamics of political patronage and political violence, both of which are rooted in colonial histories of resource extraction. Second, the Plurinational State created a host of new pathways for previously sidelined people to take on leadership roles within or alongside state entities; when cooperative miners move into these positions, they bring with them subjectivities forged in relationship to subterranean histories. Geological matters, as historicized throughout this book, have thus left their mark not only on flesh and bone but also on the hallowed halls of political and economic decision-making. The subsoil is always already present in economic, political, and social forms.

Finally, the afterword charts three "eruptions" that have emerged along the subterranean fault lines explored throughout this book. The first section examines the role of resource regionalism in the explosive end of Evo's regime in 2019, the second traces a conversation about communitarian mining that began in 2014 and continues today, and the third reflects on the proliferation of cooperative mining, illegal mining, and *jukeo* (ore theft) in the early 2020s. Overall, the afterword shows how material histories of nature and nation, as traced in preceding chapters, can help explain these contemporary eruptions. The sedimentary remains of past nationalisms do not always stay buried. Instead, they emerge through historical cracks to impinge on the present in unpredictable and often-violent ways.

SUBTERRANEAN PROPERTY:
GEOLOGY, THEOLOGY,
AND THE LAW

The 2016 conflict that ended with the murder of Rodolfo Illanes was not the first time that Bolivian mining cooperatives took to the streets in defiance of Evo Morales's government. Two years before, in March 2014, police and cooperative miners similarly confronted one another with bullets and dynamite at a highway blockade at Sayari, about sixty miles outside of Cochabamba. In this case, the object of struggle was a bill for a new comprehensive mining law, slated to be passed the following month. This law was desperately needed to replace the 1997 Código de Minería (Mining Code; Law 1777), which had been passed during the presidency of wealthy mining magnate Gonzalo Sánchez de Lozada. But mining is a controversial topic, and debates had stalled several times during Evo's first two presidential terms. The bill that finally emerged in 2014 passed easily through the House of Representatives (Cámara de Diputados) but ground to a halt in the Senate. The point of contention was Article 151, which stated that mining cooperatives could form partnerships with private companies. This, the senators argued, was unconstitutional: mining cooperatives are nonprofit entities, and this protected status could not be maintained in partnership with a for-profit company.

The Senate's decision was immediately challenged by cooperative miners, for whom the right to private-cooperative partnerships had been a core demand throughout the process of drafting the law. They pointed to other passages of the constitution to demonstrate that they should have the right of "free association" with other economic entities without jeopardizing their status as cooperatives. Yet the senators remained resolute. As the situation escalated, the country's 122,000 cooperative miners took to the highways. The standoff at Sayari left two miners dead and more than sixty people injured but did not have much impact on the new Law of Mining and Metallurgy (Law 535), which was ultimately passed on May 28, 2014. This law explicitly forbids cooperative-private alliances without supervisory involvement of the state mining corporation (COMIBOL). Seething, cooperative miners retreated to nurse their political wounds and wait for a new opportunity for engagement.

I returned to Bolivia a few months later to conduct interviews about the conflict. When I spoke with miners, they almost invariably came to our meetings armed with copies of the 2009 constitution, the old mining code, the cooperative law, and an assortment of other legal documents that they used to reconstruct their arguments. The papers were variously dogeared and crumpled, clean and reverently preserved, or covered in sticky notes and marginalia. As we spoke, I began to sense that these small booklets, available for purchase from street vendors everywhere, were the real weapon of choice for people on both sides of the conflict— as much as or more than bullets, stones, or dynamite.

The law—its formulation, application, shortcomings, and transformations—is both an everyday topic of conversation and an object of intense struggle for Bolivians. After Bolivia won its independence from Spain in 1825, its juridical framework was written by and for criollo (Spanish-descended) elites, and often the laws' primary purpose was to justify the ongoing dispossession and exploitation of the country's Indigenous majority. Transforming these laws so that they might protect the very people they were designed to exclude has been a major decolonial project of the past century—and, correspondingly, a major focus of analysis for social scientists.[1] In Bolivia the positive effects of juridical activism are evident in places as distinct as the urban marketplace, where legal protection has facilitated the economic flourishing of previously marginalized Aymara merchants, and parliamentary halls, where laws about equitable representation have ensured a level of racial and gender parity that is virtually unparalleled globally.

For an outsider documenting political conflicts, it is easy to be swept up in legal details that form the discursive rivers of everyday life. Supreme decrees and laws, each identified by numbers that are readily rattled off by even the most casual observer, often seem to be the end point of political conflicts. Legal articles, clauses, and norms are debated endlessly in public plazas and government buildings, and any attempt to represent these debates in written form quickly turns into legal soup, obfuscating as much as it reveals. In the case of the mining law, it seemed easy to forget what was actually at stake: the subterranean itself. Who can own it? Who can exploit it? Who should benefit, and whose suffering is permissible? Who makes all these decisions?

As passed in 2014, the new mining law states that the subterranean is the "property and direct dominion, indivisible and inalienable, of the Bolivian people; its administration corresponds to the State."[2] In theory, this means that the subsoil is a national inheritance, or *patrimonio*, passed along from one generation of Bolivians to another, with the state as its benevolent caretaker. In practice, the subterranean is treated as state-owned property. Three "productive actors" can apply for the right to prospect or exploit subterranean tracts: the state corporation (COMIBOL), private mining companies, and mining cooperatives. These rights are separate from land rights, even if the subsoil in question lies immediately under recognized Indigenous territory.[3] This is not particular to Bolivia—most countries worldwide have divided surface and subsurface property rights—but neither is it a "natural" form of property. As Eve Tuck has written about the division of surface and subsurface rights during the Alaska Native Claims Settlement Act of 1971, "The invention of subsurface estates is a remaking of *terra nullius*, as if somehow land a few inches below ground is uninhabited; it is a re-creation of the doctrine of discovery where there were/are already people" (2014, 252). Drawing on examples from Australia, Maria de Lourdes Melo Zurita (2020) has similarly described this as the invention of *sub terra nullius*, or an imaginary of subterranean space as somehow untouched by human activities.

Following these scholars, this chapter examines how the Bolivian subterranean was produced as a repository of state-controlled natural resources, thereby consolidating a vision of *patria* (nation) dependent on collective inheritance of subterranean *patrimonio* (patrimony). Put differently, the chapter traces an alternative genealogy of resource nationalism through the construction of subterranean resources and property. Building on the introduction, I chart a material history of resource nationalism that focuses on how subterranean spaces and materials have been constituted by human histories that both precede contemporary economic debates and continue to shape them. Although the geological stuff of the earth is significantly older than the human worlds through which it now moves, its social life has been shaped by the knowledge systems through which it was iteratively apprehended.

My focus is on how the subterranean came to be imagined as separate from the surface, or how layers of geological matter were envisioned as three-dimensional parcels that could be rented or exploited but never sold, since they collectively constituted the material basis of the nation. In Bolivian mining contracts, which grant companies or cooperatives the right to prospect or exploit subterranean resources in a particular place, the subterranean is measured in *cuadrículas* (grids), each of which is shaped like an inverted pyramid: the apex is in the center of the earth, and the base corresponds with an area of land measuring five hundred square meters, or twenty-five hectares. Even more than land, where property boundaries can be made visible on the earth as well as a map, thinking about the subterranean as *cuadrículas* with ownership, use, and access rights requires a leap of abstract faith. Yet these abstractions did not emerge, fully formed, from lawmakers' heads. Instead, I argue that they crystallized from residues left behind by earlier territorializing knowledge practices. Before it was enshrined in law, this way of seeing the subterranean had a long history that involved knowledge production on multiple continents. Accordingly, it also yoked together a set of meanings that were more-than-capitalist in origin and that continue to inform contemporary expressions of resource nationalism.[4]

The history I chart focuses on three periods: early colonial (mid-1500s), early republican (after 1825), and postrevolutionary (after 1952). During these periods, I argue, subterranean property law was constituted in conversation with forms of expertise that naturalized a nationalist interpretation of the subterranean as a shared inheritance (in which the notion of national "sharing" obscures the fact that many people had to be physically and discursively displaced from the subsoil to make it available as the inheritance of others). Specifically, I explore the shift from religious *theology*, in which the subterranean was envisioned as a God-given gift to the Spanish Crown, to scientific *geology*, in which the subterranean was envisioned as an ordered set of strata that preserved the past and yielded future wealth. This intellectual history matters because it shapes the terms of debate in contemporary legal battles, defining not just the outcome but also the contours of struggles over the subterranean. It delimits the questions that are permissible within the confines of the law (e.g., who can exploit the subterranean?) and those that are not (e.g., who makes this decision?)

The conflict around the 2014 mining law provides an entry point for this discussion, and for the book's broader arguments about the relationship between nation and the subterranean. Striking about this conflict was that at no point did anyone challenge the dominant imaginary of the subsoil as a national inheritance; instead, the dispute centered around the question of whether individual heirs could justifiably reap private profits, or whether the state ought to be involved in redistributing that income. The triad of state (*patrón*), nation (*patria*), and subterranean (*patrimonio*) remained the unchallenged grounds on which the struggle unfolded. These grounds are what I seek to historicize.

I begin by exploring the role of theology and theories of "natural law" in the establishment of subterranean property regimes during the early colonial era, when subsoil resources were claimed by the Spanish Crown even as the surface was carved up for individual conquistadors. I then use the writings of French natural historian Alcide d'Orbigny, who was hired by the nascent Bolivian state in 1830, to consider how colonial theological justifications were domesticated and secularized in the early republican period.[5] This section also goes further back historically to consider d'Orbigny's training in France, where the discipline of geology emerged entwined with environmental determinism and scientific racism—an influence that showed up in d'Orbigny's work. Next, I discuss how d'Orbigny's ideas were institutionalized within twentieth-century geological agencies, which codified the subsoil as state-administered property in the postrevolutionary era. The final section returns to the 2014 mining law to explore how the constitution of the subsoil as a national inheritance also created the possibility of national theft, a crime that is increasingly attributed to cooperative miners.

The legal separation of soil from subsoil in Bolivia predates the nation itself. This section examines how the subterranean was imagined and materialized during the colonial era, paying attention to how theories of natural law—as informed by Christian theology as well as Roman law and Aristotelian philosophy—came to underpin a regime that relegated Indigenous peoples to the surface of the earth. This history offers an empirical demonstration of Carl Schmitt's claim that "all significant concepts of the modern theory of the state are secularized theological concepts" ([1934] 1985, 36); it also shows how subterranean property laws in Bolivia—as in much of the world—were constituted by racist exclusions from the very start.

From the medieval era, Castilian interpretations of sovereign patrimony were predicated on the idea of religious right in which the sovereigns—kings and queens—drew their power from being God's earthly representatives. Inheritance rights passed from royal rulers to their offspring with Christian blessings. This approach to sovereignty was established before and during the seven-hundred-year Reconquista, in which a diverse set of Christian kingdoms on the Iberian Peninsula pushed Moors (Muslims) and Jews southward, territorializing Christianity as they went. Their primary theological system of justification was based on the categorization of all non-Christians as enemies of Christ since they had received but rejected Christianity's apostles (Wynter 2003). This cannot be understood as a project of *national* expansion, since Spain was not a unified entity until sometime after the marriage of Queen Isabella I of Castile and King Ferdinand II of Aragon in 1469. By the time the newly united forces declared victory over the Islamic Empire following the Battle of Granada in 1492, however, Spain was a nation founded through religious conquest. Religious distinction, moreover, was inextricable from the Spanish theories of genealogical descent, and in particular the Spanish obsession with *limpieza de sangre* (blood purity), which described the degree to which one could claim a supposedly pure Christian lineage. While conversion to Christianity was possible, it was Christian ancestry that granted access to power in early modern Spain. In this sense, religious difference at the time might be better understood as an early form of racial difference (de la Cadena 2005; Martínez 2004).

Throughout the Reconquista the subsoil had been a mechanism through which Isabella and Ferdinand consolidated their power in Spain.

They had been granting tracts of land to Spanish nobility and other feudal groups in order to territorialize their claim over Islamic districts, but this created a jurisdictional challenge: How could the sovereign claim ownership over and extract wealth from a territory comprising hundreds of private land claims? Historian Jeannette Graulau (2011) argues that medieval Spain consolidated its territories through the twinned principles of sovereign ownership and heavily taxed mining rights. Drawing on the precedent of Roman law, the growing empire was sliced into two vertically stacked domains: land (private, familial patrimony) and subsoil (Crown patrimony). Those who were granted mining rights had to pay a hefty percentage of the extracted metals to the imperial treasury (16). Thus, although both mining rights and taxation laws were inspired by Roman jurisprudence, they were guided by Christian notions of religious dominion incarnated in the sovereign ruler. A direct connection was drawn between the body of the sovereign and the metallic body of subterranean resources—a connection that distinguished Castilian mining law from parallel legal systems in medieval England, Germany, and the Islamic Empire, where mining at the time followed more feudal lines.

The basic distinction between sovereign ownership of the subterranean and private ownership of the land, moreover, underwrote Spanish expansion in the "new" world. Following their victory in the Battle of Granada—and suddenly at a loss for how to continue expanding their frontier—the new Spanish sovereigns agreed to finance the ambitious travel plans of the Genovese explorer Christopher Columbus. When waves of violence began rippling out from the Caribbean—northward through the Aztec Empire, through present-day California and Mexico, and southward through the Incan Empire, down the western coast of South America—Spanish rulers adapted their existing legal arsenal to ensure control over distant lands.

On the earth's surface, the Spanish followed a feudalist model in the establishment of encomiendas, in which the Crown "entrusted" a specified number of Indigenous Americans to individual Spaniards as a reward for their services during the conquest. The *encomenderos* (holders of encomiendas) were tasked with providing (unwanted) religious education, in (unjust) exchange for which they were entitled to extract tribute from Indigenous peoples in the form of labor or taxes. Although encomiendas were not formally land grants, in practice they functioned as such, since *encomenderos* had limited responsibilities to the Crown (Quispe-Agnoli 2011, 264–65). By contrast, the system of subterranean rights reinforced the Crown's sovereignty via a heavy tax regimen. Spanish rulers developed two legal mechanisms for managing the subterranean: *capitulaciones*, which were general mining rights for an indefinite period, and *asientos*, which were rights to a specific mine within a specific time horizon. In both cases, the Crown retained ownership and demanded that one-fifth of riches be delivered to the royal treasury. Like encomiendas, however, *capitulaciones* and *asientos* also conferred the right to forced Indigenous labor. All forms of abuse were unofficially permitted within the mining concession, provided that the Crown received its dues (Graulau 2011).

Both the aboveground and belowground regimes shifted after the famous Valladolid Debate of 1550, which presented a major challenge to the spiritual morality of Spain's colonial enterprise writ large. At the heart of this debate, which pitted academic theologian Juan Ginés de Sepúlveda against the Franciscan priest Bartolomé de las Casas, was the question of "natural law" in the "new" world.[6] As conceived by Thomas Aquinas in thirteenth-century Italy, the concept of natural law had been gaining popularity in Spain at the time. Aquinas, whose work aimed to synthesize key principles of Christian theology, Aristotelian philosophy, and Roman jurisprudence, argued that natural law "is natural not in the sense of being merely instinctive to man as an animal but in the sense that man, as a rational being, is naturally able to distinguish right from wrong through the use of reason" (Alves and Moreira 2013, 27). While the relationship between natural law and human law (formal jurisprudence) might vary from place to place, Aquinas suggested, certain "rational principles" would remain consistent.

This contention became the shared ground on which Sepúlveda and Las Casas staged their debate over whether Spain could ethically justify enslaving Indigenous people.[7] Insisting that the absence of a recognizable legal system in the "new" world demonstrated how far Indigenous Americans were from upholding natural law, Sepúlveda argued that Spain was morally justified in its use of force. Las Casas, meanwhile, pointed to customs and institutions that demonstrated the degree to which Indigenous Americans *did* adhere to natural law, and argued that the Spanish Crown should lead the Indians not as "beasts" but rather as "men and brothers" (Hanke 1974). The opponents' shared acceptance of natural law as a metric for measuring humanity reveals the degree to which Christian notions of nature—in which human exceptionalism was proved through a demonstration of rational thought—underwrote early legal formulations of nature, nation, and state.

Las Casas emerged victorious from this debate, which marked the end of the encomienda system. Nevertheless, his victory was fraught: although he defended Indigenous Americans against slavery, he is also remembered for having suggested that Africans could replace enslaved Indigenous Americans. Moreover, his theories laid the groundwork for a new regime of right, defending the Spanish conquest on the basis of their "care" for Indigenous people—that is, their coaxing away from "natural" impulses and toward Christian morality. In the decades after this debate, Spain granted Indigenous peoples the legal but subordinated status of "native vassals" and created for them a separate system of laws and tribunals under the Republic of Indians, juxtaposed against the dominant Republic of Spaniards. As historian Brooke Larson (2004) has demonstrated, these distinct corpuses of law not only marked the different economic and political rights and responsibilities that accrued to citizens of each republic but also divided the two groups spatially, particularly in the Andes. These divisions continued into the twentieth century despite paper commitments to liberalizing them. Theologically shaped notions of natural law, intertwined with earlier conceptions of blood purity, thus provided the foundation on which later laws pertaining to property and citizenship were erected.

Subterranean property structures were also reorganized after the Valladolid Debate. Although the legal precepts of sovereign patrimony of the mines and the one-fifth tax remained unaltered, the Crown granted "new"-world viceroys more power in enacting mining ordinances that could manage labor and capital in the absence of fully unfree labor (Graulau 2011). This was the context in which Peruvian viceroy Francisco Álvarez de Toledo, as described in the introduction, forcibly resettled Andean Indigenous peoples into 840 towns called *reducciones* (located within administrative units called *repartimientos*) that facilitated the extraction of their Crown tribute in the form of money and labor in the Potosí silver mine. As many Andean scholars have shown, the term *Indian* was solidified during this time as a tax category rather than a marker of inherent racial identity. In other words, the Spanish need for an "ethical" source of free subterranean labor cemented spatial and racial divisions between Indigenous and non-Indigenous communities (which included those of Spanish, African, and mixed descent) (Harris 1995; Stern 1993). This colonial arrangement crystallized a set of hierarchically arranged relationships to the subsoil that remain germane today: the Crown owned the subsoil, those of Spanish (Christian) descent could obtain permission to operate mines, and Indigenous peoples were confined to particular plots of land—framed as their "natural homes"—where they could be systematically forced to labor underground as tribute to the Crown. The Crown acknowledged no precolonial Indigenous engagements with the subsoil, effectively justifying its own subterranean occupation.

Although the overt authority of religious texts diminished over the next several centuries, the theologically justified segmentation of people and nature remained salient. Emerging discourses of liberalism and science did not so much revolutionize as secularize the earlier era's juridical foundations. In this context, the knowledge produced by scientists and natural historians increasingly legitimized the secular leadership of governments across Latin America, while also cementing the division between the subsoil and the surface.

GEOLOGIZING *PATRIA* IN THE REPUBLICAN ERA

In 1825, after sixteen years of war, Bolivia became the last South American country to win its independence from Spain. In the same year, a young French zoologist named Alcide d'Orbigny was commissioned by the Museum of Natural History in Paris for a three-year voyage to South America. Despite having thus far demonstrated his scientific chops only in a recently published book about mollusks, the museum decided that the twenty-four-year-old had everything needed for the voyage: youth, health, and training in the various branches of natural history (Béraud 2000). Although it was not his initial intent, d'Orbigny ended up spending the bulk of his time in Bolivia, where the newly independent state government hired him to map geological resources. The knowledge that d'Orbigny produced—which was ethnological as well as geological—preserved but secularized the separation between surface and subsoil. While indigeneity remained grafted to the surface, the subsoil was transformed from Crown property into the material basis of the new nation-state, effectively geologizing *patria* in the moment of its formation.

D'Orbigny set sail at a particularly important conjuncture in European–Latin American relations. While revolutionary passion had been ricocheting between Europe and the Americas, European philosophers had been building the foundations of modern liberalism, and liberalism was remaking how the two regions understood their shared commitments to political freedom, economic growth, and scientific reason.[8] Liberalism called for secularism, or at least a formal separation between church and state; to fight against the monarchies was to fight against their claim to divine right. Secularism, in turn, involved a reexamination of nature—both of external nature, whose contours could no longer be ascribed to intelligent design, and of human nature, which could no longer be arranged in a "natural" hierarchy from kings to commoners.[9]

D'Orbigny had already been in South America for four years when he received a letter from Andrés de Santa Cruz, president of the new Bolivian republic, inviting him to explore the country. Like other leaders of new Latin American countries, Santa Cruz needed to rebuild an economy that had been devastated by the war for independence and he thought that d'Orbigny might be able to help with the reactivation of the mining sector.[10] With this goal in mind, Santa Cruz asked d'Orbigny to look for new mercury deposits and to complete a geological map of the country, which would be the first of its kind. Santa Cruz told the young explorer that he would have the full support of the Bolivian government in his travels, in addition to the assistance of a "pair of youths" to keep him company, and financial support should he need it. D'Orbigny accepted the invitation but refused the financial support—although he did take the pair of youths, whom he intended to train so that they might continue his scientific collection after his departure. He also requested pack animals and "some indios" to drive the animals.[11] Santa Cruz fulfilled both requests. When d'Orbigny finally took off, he did so with a full team of sixty Indigenous people who rowed boats, carried supplies and samples, and directed the naturalist around the country (Mendoza L. [1971] 2002, 233).[12]

During his lifetime nearly all of d'Orbigny's publications were written in French, and in Bolivia they were accessible only to a small community of elites (Albarracín Millán 2002). The most famous of his works were eleven volumes of travel diaries, *Voyage dans l'Amérique méridionale* (Voyage in southern America, 1835–47), but he also wrote an ethnological treatise titled *L'homme américain* (The American man, 1839) that gained some popularity in France. The only book he published in Spanish was *Descripción geográfica, histórica y estadística de Bolivia* (Geographic, historical, and statistical description of Bolivia, 1845), which was written at the explicit request of then president General José Ballivián (Quesada Elias 1991). Nothing else was translated into Spanish until 1907, when Victor E. Marchant Y., section leader of the Ministry of Colonization and Agriculture, collected geologically oriented sections of d'Orbigny's *Voyage* and published them as a standalone book called *Estudios sobre la geología de Bolivia* (Studies on the geology of Bolivia; d'Orbigny and Marchant Y. 1907). Tucked into the back pages of this book was the country's first geological map, with apologies from the editor for its reduced size. In 1944 an Argentine press translated *L'homme américain* into Spanish, and in 1946 the Anglo-Bolivian Cultural Institute reprinted *Descripción* in an extended two-volume version.

All of this is to say that, despite having been written in the early nineteenth century, d'Orbigny's work became widely accessible to Bolivians only in the first half of the twentieth century. In analyzing d'Orbigny's work, therefore, one must attend to the scientific context in which d'Orbigny was formed, the actual contents of his writings, and the political context in which his ideas were taken up. I focus primarily on the first two points in this section, but I return to the last one later in this chapter.

When d'Orbigny set sail at the turn of the nineteenth century, the discipline of geology was still in its infancy—British naturalist Charles Lyell would not write the first volume of his definitive treatises on the subject until 1830—but the debates that would frame the discipline were well underway. Paris was the center of the emerging fields of natural science, and the new Museum of Natural History (the postrevolutionary name given to the Royal Botanical Gardens) was the center of Parisian intellectual life. While studying at the museum, d'Orbigny had two influential mentors: Georges Cuvier, a comparative anatomist remembered for his interlocking contributions to biology, geology, and scientific racism; and Alexander von Humboldt, whose travel accounts d'Orbigny read with reverence. In different ways, both these men established a preference for "seeing" landscapes vertically, as well as for naturalizing the presence of particular groups of people in particular places. Yet they had distinct methodologies that showed up in d'Orbigny's work as a kind of doubled gaze. On the one hand, d'Orbigny practiced a kind of visual dissection of people and rocks that closely resembled Cuvier's (physical) dissections, but on the other hand, d'Orbigny extolled the beauty of the landscapes through which he traveled with a Humboldtian romanticism. Although these gazes were not as clearly divided by subject matter in d'Orbigny's original (French) publications, in subsequent Spanish translations the former tendency (dissections after Cuvier) was more clearly seen in geological writing, while the latter tendency (landscapes after Humboldt) was muted—only to reappear in twentieth-century nationalist writings about landscape and national character. For this reason, it is worth examining the racist thinking that drove Cuvier's and even Humboldt's work and that was ultimately naturalized in subterranean property law.

When Cuvier joined the Museum of Natural History in Paris in 1795, there were at least three competing groups of scientists claiming rocks as their object of inquiry: physical geographers, who mapped morphological features visible on the surface of the earth; mineralogists, who collected and cataloged types of rocks and rock formations; and "geognocists," who had developed a technique for depicting the earth in "stratigraphic columns" that resembled slices of a many-layered cake.[13] For Cuvier, the limitation of geognosy was that it failed to provide a historical explanation for the development of rocks. With this deficiency in mind, he combined the visual methods of geognosy with his own methods of comparative anatomy.[14] Drawing data from the quarries surrounding Paris, Cuvier used the observation that the fossilized remains of different creatures were present in different strata to support his scandalous belief in species extinction, which he claimed had been driven by catastrophic environmental events.[15] This argument was the basis for Cuvier's regionalist approach to natural history, since floods and volcanic eruptions would not have affected every region of the planet equally. Regionalism, in turn, laid the groundwork for Cuvier's growing interest in human racial "types," which he understood as having emerged through spatial separation brought about by geological events. In line with his catastrophe-based approach to natural history, he claimed that humans had escaped to different corners of the earth following the last natural disaster (about five thousand years prior), where they had developed in isolation into three separate—and hierarchically distinct—races. "Caucasians," he claimed, had remained unchanged from biblical times, whereas "Mongolians" and "Ethiopians" had both degenerated in unfavorable climatic conditions (J. Jackson and Weidman 2004).

In his examination of change over time—changing geological formations and changing human races—Cuvier contributed to prevailing theories of environmental determinism, in which human type was determined by regional environmental features. However, by employing dissection as his primary tool in comparative anatomy, he also contributed to an emerging set of discourses that biologized rather than environmentalized racial difference. These discourses, now known as *scientific racism*, located racial difference in internal rather than externally visible bodily features.[16] Thus, Cuvier's geology emerged already entangled with environmental determinism and scientific racism well before d'Orbigny's voyage to Bolivia.

Cuvier's influence is most clearly visible in d'Orbigny's book *L'homme américain* (1839). This is also the book most often forgotten by those who celebrate d'Orbigny's legacy, since it lays out a troubling series of Indigenous racial types. While the book is dedicated to Humboldt, the introduction makes clear that the project was inspired by Cuvier. Since Cuvier had not himself traveled to the Americas, he "had not seen fit" to include Indigenous peoples within any of his "three great races," and d'Orbigny notes his desire to continue Cuvier's efforts (viii–ix). Over the course of his eight years in South America, d'Orbigny spent time with Guaraní, Mapuche, Aymara, and Quechua people, among others, and in *L'homme américain* he categorizes them into three "races" based on shared physical traits ("Ando-Péruvienne," "Pampéenne," and "Brasilio-Guaranienne"), seven "branches" based on shared geographies, thirty-nine "nations" based on shared root language, and dozens of "tribes" based on shared dialects (see figure 1.1). As anthropologist Deborah Poole describes it, these groupings appear as "an amalgam of linguistic and physiological characteristics—as if d'Orbigny did, indeed, wish to merge the study of bodies and words into a single particularizing grid for the comparison and classification of humans" (1997, 79).

Like Cuvier's, d'Orbigny's attention to detail was meticulous and dismembering. He compared Indigenous peoples in terms of facial features (noses, chins, jaws, foreheads), height, skin color, and even smell. He communicated all this information through typological tables, creating neat grids through which to render individuals and collectives legible to colonial eyes. Although he was not literally dissecting in the style of Cuvier, he sliced bodies in a way that conjures dismemberment. For example, he described Indigenous women thus: "They are too robust, too wide to be well made, in the sense that we give this word in Europe. Nature has endowed them, on the other hand, with all the desirable advantages for the kind of existence that calls them: broad shoulders, modest breasts, well-proportioned throat . . . wide hips; also the act of childbirth is always easy and never has unpleasant consequences; small hands and feet" (108–9). Such bodily segmentation and explicit comparison to a European ideal of feminine beauty (understood as the opposite of utility) is epistemologically violent; the focus on the body's capacity for labor and biological reproduction without "unpleasant consequences" sounds more

RACES.	RAMEAUX.	NOMS DES NATIONS.	LIMITES D'HABITATION DES NATIONS: en latitude australe.	en longitude ouest de Paris.
1.re Race. ANDO-PÉRUVIENNE.	1.er Rameau. PÉRUVIEN...	Quichua ou Inca.	0° au 28°	65° au 83°
		Aymara........	15° au 20°	69° au 75°
		Chango........	22° au 24°	72° 30'
		Atacama.......	19° au 22°	72° 30'
	2.e Rameau. ANTISIEN....	Yuracarès......	16° au 17°	66° au 69°
		Mocéténès.....	16° =	69° au 71°
		Tacana........	13° au 15°	70° au 71°
		Maropa	13° 50'	70° =
		Apolista.......	15° =	71° =
	3.e Rameau. ARAUCANIEN.	Auca ou Araucano	30° au 50°	60° au 76°
		Fuégien........	50° au 56°	68° au 77°
		Patagon ou Té-huelche......	39° au 53°	65° au 74°
2.e Race. PAMPÉENNE.	1.er Rameau. PAMPÉEN....	Puelche........	34° au 41°	60° au 68°
		Charrua.......	31° au 35°	56° au 62°
		Mbocobi ou Toba	21° au 32°	61° au 64°
		Mataguayo.....	22° au 28°	63° au 65°
		Abipones.......	28° au 30'	61° au 64°
		Lengua	27° =	62° =
	2.e Rameau. CHIQUITÉEN..	Samucu........	18° au 20°	60° au 62°
		Chiquito	16° au 18°	60° au 64°
		Saravéca.......	16° =	62° =
		Otukè	17° =	60° =
		Curuminaca	16° =	62° =
		Covaréca.......	17° =	61° =
		Curavès.......	19' =	60° =
		Tapiis..........	18° =	60° =
		Curucanéca.....	16° =	62° =
		Païconéca	16° =	63° au 64°
		Corabéca.......	18° =	62° =
	3.e Rameau. MOXÉEN....	Moxos	13° au 16°	64° au 69°
		Chapacura......	15° =	64° au 65°
		Itonama........	13° au 14°	65° au 67°
		Canichana......	13° au 14°	67° au 68°
		Movima........	14° =	68° au 69°
		Cayuvava......	12° au 13°	68° =
		Pacaguara.....	10° =	67° au 68°
		Iténès.........	12° au 13°	67° au 68°
3.e Race. BRASILIO-GUARANIENNE....		Guarani........	du 34° de l. a. au 14° de l. b.	37° au 64°
		Botocudo.......	18° au 20°	43° =

FIGURE 1.1. D'Orbigny's racial taxonomy. From *L'homme americain* (1839, 11).

like a description of an animal than a person. Just as Cuvier physically sliced human and animal bodies, so did d'Orbigny descriptively slice generalized "Indian women."

D'Orbigny also sliced cartographically. In his typological table, he delimits each "Indian nation" within longitudes and latitudes, effectively turning the landscape into a series of interlocking puzzle pieces that each contained a single group of people. These territorial limitations, he argued, affected racial development indirectly. Unlike his contemporary environmental determinists, he believed that humidity levels rather than temperature caused differences in skin color. Without an instrument to measure humidity, however, he judged it based on whether the plants he pressed in his notebook dried or rotted (78), often using elevation as a proxy. In practice, this way of discussing indigeneity worked to fix communities along the surface of the earth, bounded not only in two dimensions (longitude and latitude) but also by altitude. That these territorial boundaries were not fully "natural"—and that they had been at least partially generated by colonial reorganization schemes such as Toledo's *reducciones*—was not considered in d'Orbigny's notebooks.

On its face, d'Orbigny's attention to landscape and environmental milieu more generally was a practice passed along from Humboldt rather than Cuvier, since the latter believed that life was shaped from within rather than from the outside. Unlike Cuvier, Humboldt was a traveler—a field scientist rather than a laboratory scientist—and the five years he spent traveling in the Americas at the turn of the nineteenth century were crucial to his scientific project. Instead of Cuvier's scalpels, microscopes, and specimen trays, Humboldt's scientific instruments included compasses and dipping needles to measure the earth's magnetic field, barometers to measure altitude, hygrometers to measure humidity, and even a "cyanometer" to measure the blue of the sky (Walls 2009, 38–39). In this way, Humboldt's gaze was decidedly physiognomic, meaning that it demanded a mobile, observing subject who could perceive features of the landscape from various angles and create an impression of the whole (Poole 1997). Although he took pages of detailed measurements like any good Enlightenment scientist, Humboldt's visual and literary representations were inspired by German romanticism and conveyed a near-religious sense of beauty and awe (Walls 2009).[17]

Yet Humboldt's scientific enterprise shared some commonalities with Cuvier's research agenda. Like Cuvier, Humboldt borrowed the techniques of geognosy to develop his interests in the relationship between rocks and life. Humboldt's passion was botany, but he had spent the early years of his adult life at the Freiberg School of Mines in present-day Germany, where he was trained in the principles of geognosy and mineralogy. There, he spent four to five hours underground every day and became particularly fascinated with "subterraneous vegetables," especially "cryptograms," a group of plants that included mosses, lichens, and fungi. This interest was partly practical: Humboldt had been tasked with studying the effects of this flora on miners' respiratory health. But as historian Patrick Anthony (2018) has demonstrated, it was also here, underground, that Humboldt developed his fascination with the question of how to represent historical plant migration. He used geognostic profiles to show how plant species varied with depth; later, working aboveground in South

FIGURE 1.2. Humboldt's *Tableau physique des Andes et pays voisins* (Humboldt and Bonpland 1807).

America, he used this same technique to show how plant species varied with altitude. In Humboldt's most famous painting, which illustrates botanical variation along the folds of Mount Chimborazo, the influence of geognostic stacks is clearly visible (see figure 1.2). In its carefully delineated rows of botanical zones that correspond with different elevations, Mount Chimborazo is like a mine turned inside out. Ancient mining techniques became new tools in Humboldt's appraising eyes, allowing him—as he later boasted—to represent "entire countries as one would a mine" (quoted in Anthony 2018, 33).

Humboldt's descriptions might have been romantic, but he was visually dissecting nonetheless—like Cuvier but on a grander scale. In d'Orbigny's work this tendency is most evident in *Estudios sobre la geología de Bolivia* (d'Orbigny and Marchant Y. 1907), which—unlike *L'homme américain*—is an edited compilation from d'Orbigny's travel logs. Much like Humboldt's famous travel diary that d'Orbigny had pored over prior to his journey, *Geología* reads like a personal account of a journey across space. All the descriptions are in the first person, unfolding the landscape from d'Orbigny's ambulatory point of view. To offer two examples:

> I crossed this chain [of mountains] perpendicularly to its orientation, [and] here is what I saw: the first branch is, as I said, a little less elevated than the second, composed of rich layers of quartzose sandstone, often very friable, with reddish coloration. I have not seen any fossils, but many traces of copper, either in oxides, or as infiltrate between the layers, or disseminated in tongues. (9)

> I crossed the slates of the Silurian period to the Pampa de Ruis, a small sort of valley, located in the middle of the hill. There the accumulated layers of slate rocks end, and the Devonian sandstones begin to be found again. I ascended . . . to the top of the hill, where a vast plateau, dominated by sandstone mammoths in almost horizontal layers, forms something like a chain of low elevation. (169)

Although these descriptions are told from d'Orbigny's point of view, they are remarkably impersonal. In the first quote, his gaze is down, toward the rocks, and he is dividing and classifying their mineral structure as he goes. In the second quote, he narrates his ascent in relation to the geological origins of the rocks, effectively moving through time as well as space. By the time he sees the whole landscape from the top of the hill, he has already temporally cataloged its individual parts. A continuous description of what he saw, d'Orbigny's geological writings thus shift from the layers visible only up close, which are described with objective disinterest, to the landscape, which is often described with poetic language that would have appealed to Humboldt's romantic sensibilities.

Even though d'Orbigny approaches the study of both people and rocks with the same gaze, people are as absent from d'Orbigny's geological writings as geology is from his studies of people. The only reference to people in *Geología* is the note that "Indians wash . . . the sands of the river" in search of gold (13). In part, this is a product of the fact that *Geología* was a compilation of geologically focused chapters from d'Orbigny's travel diaries, curated by the translator. But this compilation was actually a simple project for the translator since, although the diaries as a whole were organized by geographic region, each region had subsections that were easily collected by the twentieth-century editor. In either case, it is striking how much *L'homme américain* refers to landscape, elevation, and climate, closely associating Indigenous peoples with their (surface-level) environment, while subterranean rocks are decidedly *not* associated with the people above them. Given the brief note that people were actively looking for gold, it is also striking that this is not included in *L'homme américain* in the list of Indigenous livelihoods, which are instead identified as hunting, agriculture, and pastoralism. Subterranean geology is entirely free of cultural particularities and ready to be territorialized otherwise: a true *sub terra nullius*, indeed.

The separation of environmental from cultural data points was not unusual to d'Orbigny or even geologists of Latin America. For instance, Bruce Braun (2002, 52–53) demonstrates how George Mercer Dawson, a geologist working in western Canada in the 1870s–1880s, divided his notebooks down the center of each page, so that he could record geological observations in one column and cultural observations in the other column, a division that effectively pried a single place into two components. Braun goes on to describe how this separation legitimized the spatial incarceration of Indigenous peoples within reservations. What I want to underscore is that this spatial incarceration was three-dimensional. Although d'Orbigny attended closely to elevation in discussing Indigenous peoples, his thinking was literally shallow: the subterranean is absent from *L'homme américain*.

The perceived importance of *Geología* is evident in the book's introduction, written by Bolivian archaeologist and geographer Belisario Díaz Romero, who declared his expectation that it would be "useful to the diverse industries that now are starting to exploit the Bolivian soil" (1907, xviii), a group in which he enthusiastically included miners, agriculturalists, and geologists. Nearly a century later, Bolivian sociologist Juan Albarracín Millán would similarly observe that "without [d'Orbigny's writings], there could not exist in Bolivia any possibility of developing a modern national culture based on scientific knowledge of the Bolivian reality" (2002, 282). Both men glide past d'Orbigny's ethnographic work, an absence made possible by the compartmentalization of d'Orbigny's observations: people corralled within identifiable squares of land, and the subsoil as a repository of historical treasures divorced from activities on the surface.

The newly independent Bolivian state used d'Orbigny's geological knowledge to engulf local land claims, since any human land claim was both temporally preceded and physically undergirded by geological history. Geology did not make the primordial cut that separated land from subsoil, but it reinforced in secular terms what Spanish legal scholars and theologians had already worked to achieve several hundred years prior. This geological vision of the subsoil as abstract space, easily divided into three-dimensional pyramids of rock to be rented like state-owned basement apartments, still undergirds contemporary property law. This way of seeing was not fully consolidated in d'Orbigny's time, but the long process of translating, reassembling, and institutionalizing his oeuvre—a process that took place in the twentieth century—slowly naturalized the separation between land and subsoil in Bolivian *patria*.

DEPOLITICIZING THE SUBSOIL IN THE TWENTIETH CENTURY

In contemporary La Paz, a group of Bolivia's most successful retired geologists gather every Wednesday afternoon in a chic Western-style café in the southern zone of the city, where wealthy, light-skinned Bolivians have historically congregated. Of this group of around ten geologists, some are still working as consultants, many spend their days golfing together, a couple are hard of hearing, and all are *q'ara* (Quechua for white or non-Indigenous) men. Their ranks include a former president of COMIBOL, a former minister of mining and metallurgy, and an elderly Scotsman—the only foreigner among them—who claims to have single-handedly discovered the San Cristóbal silver deposit, currently the largest mine in Bolivia.

When I joined this group one Wednesday in February 2017, I received a warm welcome that was likely conditioned by expectations about what a gringa geographer would be interested in knowing. Once I had explained to them that I was interested in learning about the relationship between geological exploration and politics in twentieth-century Bolivia, however, they looked perplexed. As I switched on my audio recorder to begin an impromptu focus group, Manuel, a recent president of the College of Geologists, summed up the group's general puzzlement:

> Geology is the study of nature. It's never been political, it's a pure science. . . . We are like doctors for the earth. [Medical] doctors diagnose what you have in your body, and the geologist does the same: he deduces what has happened in the earth. The form the work takes is drilling; just as a doctor takes a sample of your blood, geologists take a sample of rock. . . . If there's a political component, it's that there's never been enough support for our work. There's no

political mining framework [*política minera*] to apply geology as an instrument of progress. (La Paz, February 17, 2017)

Manuel went on to tell a comical story about his grandmother, who had misheard his adolescent announcement that he would study *geology* as an intention to study *theology* and had expressed delight that they would finally have a priest in the family. Once corrected, she questioned his decision, as she had never heard of anyone studying geology. In this way, Manuel suggested, she was a typical Bolivian: few Bolivians truly understood the value of geological exploration. The Bolivian state exemplified this tendency, having repeatedly failed to center geological mapping and prospecting in its economic development plan since the nationalization of the tin mines in 1952.

Yet as the interview progressed, I realized that Manuel's stance was challenged by evidence from his own life story. As an undergraduate at a state university in the decade immediately after the 1952 National Revolution, he and his fellow students had been tasked with producing geological sheets of unmapped parts of Bolivia's subsoil, with the idea that the work of enough students could create a detailed map of the national underground. His sheet won him a job on the first team of geologists to form DENAGEO (Departamento Nacional de Geología; National Department of Geology), the nation's earliest independent department of geological study. As I elaborate in the following, the formation of DENAGEO in 1960 marked a contentious moment of centralizing geological knowledge. But Manuel saw this all as entirely unpolitical mapmaking geared toward national progress—a goal that was never achieved due to a lack of government support.

Manuel's conviction that earth sciences are separate from political processes is far from unusual. It speaks to the consolidation of the subsoil as purely a natural realm, protected from any contaminating influences of society and politics. Of course, his grandmother's inadvertent mistaking of geology for theology is extremely apropos: the subsoil, once conferred on the Crown by divine right, as bolstered by theological knowledge, is now conferred on the state by juridical right, as bolstered by scientific knowledge. Manuel's life history actually demonstrates how this happened in the years after the 1952 National Revolution, when geological knowledge was increasingly institutionalized within the state. This process of postrevolutionary institutionalization is my focus in this section.

D'Orbigny's publications, finally translated into Spanish in the early twentieth century, were crucial to the processes of both politicizing national land in the 1930s–1950s and depoliticizing the subsoil in subsequent decades. In the earlier period, historians and geographers harnessed d'Orbigny's writings on people and their surface-level environments to territorialize a mestizo national identity. For instance, some of the first inklings of a mestizo resource nationalism are apparent in the writings of Jaime Mendoza, an early twentieth-century physician, novelist, geographer, and avid reader of d'Orbigny.[18] Mendoza was writing in conversation with then-prominent eugenicist Alcides Arguedas, whose book *Pueblo enfermo* ([1909] 2008) had argued that Bolivian geography had negatively shaped the character of Indigenous peoples and was obstructing national development (Sanjinés C. 2004, 46). Drawing on d'Orbigny, Mendoza flipped Arguedas's theories on their head: in *El factor geográfico en la nacionalidad boliviana* (The geographic factor in Bolivian nationality, [1925] 2016) and *El macizo boliviana* (The Bolivian massif, [1933] 2016), Mendoza argued that the formidable Andean mountain range had positively informed the character of the Bolivian people. Mendoza was popular among nationalists because he "used geography—and the heights of the Andes and the hegemonic and administrative impulses that they inspire—to affirm Bolivia's right to exist as an independent nation" (Sanjinés C. 2004, 83). Indeed, Mendoza described the landscape as the ultimate mother of the nation, and he characterized the two Creole generals who led the revolutionary war in Bolivia, Simón Bolívar and Antonio José de Sucre, as doctors who helped deliver a nation that "still had not been born but already existed" ([1925] 2016, 283). Continuing a European tradition of gendering the nation as a woman who gestates both bountiful natural resources and "sons of the soil," Mendoza used the landscape to claim indigeneity as the mother of a mestizo Bolivian nation that had been made possible by male Creole doctors.[19] In other words, he affirmed the "natural" connection that d'Orbigny had described between Indigenous peoples and land and used it as genealogical justification for a mestizo nationhood that was coterminous with state territory.

Mendoza's fascination with the romance of elevation was taken up by later nationalist writers, including Carlos Montenegro, a journalist who helped found the political party MNR (National Revolutionary Movement), which won power after the 1952 revolution. Montenegro's book *Nacionalismo y coloniaje* ([1944] 1984) narrated the history of Bolivia as a continuous battle between the nation and the "antination," a force composed first of Spanish colonizers and then of feudal landholders. Like the "cosmic race" of Mexican nationalist José Vasconcelos ([1925] 1979), Montenegro developed a theory in which Bolivian mestizos were framed as the true national subjects, while indigeneity was located in a glorious past (Rivera Cusicanqui 2003). Even though he was working with primarily journalistic sources, Montenegro's history played out in clearly Bolivian landscapes, with frequent references to mountains, forests, and Lake Titicaca naturalizing his national history within the state's territorial borders.

Montenegro's distinction between nation and antination became crucial in the years following the 1952 National Revolution, when a homologous connection was drawn among the state (*patrón*), the subsoil (*patrimonio*), and the people (*patria*).[20] Postrevolutionary president Víctor Paz Estenssoro worked to frame the MNR's takeover of the state as the concluding chapter of Bolivia's national history: if the revolution of 1825 had signified *political* independence, he argued, then the revolution of 1952 meant *economic* independence (Nicolas and Condori 2014, 32). This framing justified state management of the subsoil as the ultimate expression of sovereignty. Once possessed by the Spanish Crown and then controlled by Bolivian elites, the subsoil was finally the property of the people. And yet who counted as "the people" remained circumscribed to those who identified primarily with the (still colonial) nation.

Meanwhile, geologists were increasingly emulating d'Orbigny's tendency to partition and categorize the subterranean without any reference to the people who lived above. By the early twentieth century, geologists had moved away from description and toward dissection and categorization; landscapes and cultural practices were left for novelists and political theorists like Mendoza and Montenegro. Whereas d'Orbigny had complemented his geognostic profiles and grid-like typologies with hundreds of pages of first-person description, twentieth-century geologists seemed to have removed themselves from the work entirely. They relied nearly exclusively on the impossible perspective: a slice of cleanly ordered vertical layers that could be apprehended all at once.

In Bolivia this shift is exemplified by the prominent work of Federico Ahlfeld. A German geologist who immigrated to Bolivia in 1924, Ahlfeld took a special interest in Llallagua's tin mines, publishing several journal articles specifically about the region (e.g., Ahlfeld 1931, 1936, 1941). However, Ahlfeld's most important contribution was the compendium *Geología de Bolivia* (Geology of Bolivia, 1946, reprinted in 1960 and 1972), which is still regularly cited. The first sentence of this book recognizes d'Orbigny's legacy, but the book is written and organized exceedingly differently than d'Orbigny's naturalist approach. Unlike d'Orbigny's eleven-volume description of his travels, Ahlfeld's book is concise, less than two hundred pages even in the extended 1972 version. The chapters are organized by geological epoch rather than geographic region, beginning with the Precambrian and ending with the Holocene (marked by the appearance of humans). In other words, Ahlfeld orders Bolivia by time rather than space. His own gaze is not directly referenced; there is no "I" in his study, even in the foreword, in which he notes that "we" felt compelled to republish the book given the absence of comprehensive data available about Bolivian geology. Most important, the book is studded with folding maps that depict history vertically, each layer textured with a different crosshatching to identify its geological epoch. These simple maps do more analytic work than any of Ahlfeld's words, which contrasts sharply with d'Orbigny's flowery descriptions and only occasional visuals. This depersonalized way of seeing the subsoil, a stripped-down version of d'Orbigny's empiricism, is precisely what underlies contemporary mining law.

The arc of Ahlfeld's career coincided with the centralization of geological knowledge within the state, increasingly beyond the control of "the people." Prior to the nationalization of the tin-mining sector, Ahlfeld had held positions as the chief geologist of the Bolivian government (1935–36) and as the chief geologist of the Directorate General of Mines and Petroleum (1935–36 and 1938–46), and he was one of two geologists on the Bolivian side during the first formal involvement of the US Geological Survey in the country from 1940 to 1941 (Kiilsgaard et al. 1992). In the years immediately prior to the 1952 National Revolution, he was actually employed at Simón I. Patiño's tin mine in Llallagua-Uncía, where I conducted the majority of the ethnographic work that appears in the rest of this book (Redwood 2003). His job vanished with the 1952 passage of DS (Decreto Supremo; Supreme Decree) 3196, which nationalized the tin mines and inaugurated the new state mining company, COMIBOL. This decree stated that COMIBOL would take charge of "exploring, exploiting, and making use of [*beneficiar*] the minerals of the mining deposits that the Government of the Nation assigns it," as well as commercializing metals and importing needed supplies.[21] In other words, COMIBOL was to centralize the work of mining—not only exploitation but also exploration and prospecting. Ahlfeld, his work associated with the private industry that was now shunned, left the country.

Geological knowledge, however, quickly became an arena in which the postrevolutionary state collaborated with the United States to limit the power of the workers' unions, as well as that of "the people" more broadly.[22] Ahlfeld's career is again instructive. He returned to Bolivia to work as a geological adviser for the United Nations (1956–60) and the German Geological Mission (1959–63). In consultation with these entities, as well as the US Geological Survey, in 1960 Bolivia established a quasi-autonomous geological agency, DENAGEO, with a mandate to create an official geological map of Bolivia (Kiilsgaard et al. 1992). This was where Manuel, with whom I opened this section, had his first job. Following the military coup that ousted the MNR in 1964, the newly empowered General René Barrientos Ortuño rechristened DENAGEO as GEOBOL (Servicio Geológico de Bolivia; Geological Service of Bolivia). A fully autonomous institution, GEOBOL was responsible not just for producing a national map but also for coordinating technical services for private companies and foreign geological agencies.[23] According to Edgar Ramírez, former director of the COMIBOL Archives in El Alto, the separation of geological knowledge from COMIBOL, first through DENAGEO and then through GEOBOL, was part of a more general process of denationalizing the mines that began before COMIBOL even got off the ground (interview, El Alto, February 22, 2016). This move fortified the state's legal control over the subsoil by wresting power from the miners' unions, which still had veto power within COMIBOL. Through centralizing this knowledge, the state territorialized the vertical depths of the subsoil, constructing the rocks below as state-owned resources.

Despite their differences, d'Orbigny and Ahlfeld were both European geologists employed for at least part of their lives by the Bolivian state, where they produced knowledge that transformed the subterranean not only into natural resources but also into state territory, devoid of the particularizing influence of local cultures. Although d'Orbigny's geological writing relied on perspectival description, he nevertheless tended to compartmentalize rock types in a way that he had inherited from Cuvier and that is visible in Ahlfeld's later work. D'Orbigny took this same approach to his study of people, but he also fixed them to the surface of the earth in a way that allowed Ahlfeld and others to study the subsoil without discussing people at all. Together, and alongside many similar scientists, they produced the kind of scientific gaze that allowed the state to consolidate national subterranean territory. With geological knowledge comes power: the skeletal frame of the state's juridical hold on the subterranean.

D'Orbigny's research secularized a preexisting, theologically justified distinction between land and subsoil, but twentieth-century Bolivian readings of d'Orbigny by nationalists and geologists deepened that divide. The imaginary of national land, as described by writers like Mendoza and Montenegro, was linked to the development of a sense of national unity—though that unity was always premised on the erasure of the *indio* (this time through assimilation rather than segregation). The subterranean, meanwhile, was constructed as politically neutral at precisely the same time that geological knowledge was being centralized and institutionalized in the Bolivian state. The treasures of the subsoil might be a collective national inheritance, but the state would be in charge of determining how best to use them.

INHERITANCE AND THEFT

If the subsoil has been constituted as a national inheritance held in trust by the state, how does this material history shape conflicts around mining laws? This concluding section returns to the 2014 mining law conflict to discuss how the naturalization of subsoil as the material foundation of *patria* was accompanied by the specter of national theft. Along with more obvious perpetrators such as imperial powers and multinational companies, mining cooperatives are increasingly figured as "enemies within" that threaten the coherent unity state, nation, and subsoil.

The backstory of the 2014 conflict was recounted to me by Héctor Córdova, a politically savvy metallurgist who held top positions in both the Ministry of Mining and Metallurgy and COMIBOL during Evo Morales's time in office. I interviewed him in June 2017 in his house in La Paz, where he prepared us tea and buttered bread before settling down in his sunny dining room. He began the story in 2010, with the convening of a three-day workshop to draft a new mining bill. This workshop, Héctor explained, had brought together everyone involved in or directly affected by mining, including representatives from CONAMAQ (the highland Indigenous federation), CSUTCB (Confederación Sindical Única de Trabajadores Campesinos de Bolivia; Unified Syndical Confederation of Peasant Workers), and workers' unions. After it was written, the only sector that remained dissatisfied was the mining cooperatives, who complained to the president directly. Evo, seeking to appease the sometimes-irascible miners, tossed out the draft and set up a new commission that only involved private mining entrepreneurs, salaried miners' unions, and—of course—mining cooperatives. According to Héctor, it was not until the newly drafted bill had been sent to the Legislative Assembly in 2014 that Evo realized he had made a mistake: "Morales realized that he was being deceived [by the cooperatives]. He gave the assembly freedom to change Article 151, because that article authorized cooperatives to associate with private companies while maintaining their status, robbing the country, because that was what [the cooperatives] wanted legally" (La Paz, June 15, 2017).

Making a deal with a private company, Héctor implied, was akin to robbing the country, since subterranean resource wealth was everyone's inheritance. Deals struck outside of the state's purview were inherently working against the nation. Moreover, cooperative miners wanted to retain their preferential taxation regime even after forming partnerships with private companies. Legally categorized as nonprofit entities, cooperative miners are not required to pay taxes. Although they do pay royalties, these are minimal and remain in their department of origin (85 percent goes to the departmental government and 15 percent to the municipal government).[24] The central state does not profit from cooperative mining activity, meaning that this wealth cannot be redistributed in the form of government-funded projects in other parts of the country. As if to add insult to injury, cooperative miners wanted to retain this regime indefinitely. Mining contracts last thirty years, at which point they can be renewed, but *cooperativistas* argued for a limitless term. In Héctor's analysis, the thieves were trying to guarantee their theft in perpetuity.

Moreover, while the mining cooperatives lost their battle in 2014, Héctor claimed that tensions continued to bubble under the surface and led directly to the 2016 conflict and the murder of Rodolfo Illanes. He elaborated, "They [cooperative miners] wanted to be allowed to associate with companies and continue earning money without doing anything. Because those cooperative miners [that form partnerships] no longer work, only the company does, and they pay the cooperatives a monthly amount—*that* is why they wanted them [partnerships] to be legalized once again." In short, Illanes died because mining cooperatives wanted to legalize their subterranean theft, and they wanted to do so with minimal personal labor. By charging a private company to operate within their contractually held areas, mining cooperatives would be acting like land-lords rather than workers—illegally subletting property that belonged to the state and robbing the people of taxes that were their birthright.

At the heart of this struggle was a question about the meaning of national inheritance. If the subsoil is a national inheritance, then does that mean that any individual citizen can profit from it, or must that benefit be redistributed by the state? While mining cooperatives adopted the former stance, Héctor and many others suggested that excluding the state and failing to redistribute profit was akin to theft. The material history offered in this chapter, however, suggests that the construction underlying *both* options relies on a particular way of seeing the subsoil as separate from the soil above it, and that this separation was forged through a long history of colonial and capitalist extractivism. Theology in the colonial era and geology in the republican and nationalist eras have both worked to naturalize a legal separation between soil and subsoil that was, ultimately, a colonial separation. The production of *sub terra nullius*—a subterranean owned by no one—has been constitutive of the modern Bolivian state and its legal system. While ultimately the landlord of the subterranean, the state is figured as a paternal caretaker of a nationally shared subterranean inheritance. The state's claim on the subsoil is thus maintained not only by the language of the law, which recapitulates the Spanish Crown's assertions in nontheological terms, but also in a massive body of geological knowledge that, stripped of its social referents, conjures an imaginary of the subsoil as abstract space, a treasure chest of resources without history.

This relationship is key to understanding contemporary invocations of *patria* in Bolivia. According to this imaginary, those who make a more direct claim on the subsoil—whether because their land lies above it or because they have labored directly in it—are common criminals. The most notorious such thieves of *patria* are cooperative miners, and it is to these complicated figures that I now turn.

2

Llallagua is a vertical-feeling town, even when you are not going underground. The main street ascends at a precipitous angle from the bus station all the way up through the central plaza and beyond, eventually connecting to the road that winds up to Cancañiri, the highest (and most frequented) mine shaft in town. In some places, the sidewalk is just a very long staircase. Along the way, men in sweaters and women dressed *de polleras* (in traditional pleated skirts) sit on stoops to visit, peel potatoes, and watch the street activity. Through open doors more people can be glimpsed eating steaming bowls of soup or sharing crates of beer.

All these people looked up when I passed by, likely noting that I was making my third pilgrimage in as many days to the offices of FERECOMINORPO (the Regional Federation of Mining Cooperatives of Northern Potosí). The FERECOMINORPO building was halfway up the main street, high enough that I had to pause to catch my breath before taking the final, extra-large step up to the entrance. Through the open door, I called out a wheezing greeting to Veronica, the young secretary, and Doña Amalia, the regional representative of women cooperative miners. Seated behind their respective desks, both women were dressed in thick winter coats with scarves pulled up to their noses. Although I was sweating from my uphill climb in the sun, their windowless front office retained the nighttime chill. They beckoned me indoors to cool off and catch my breath.

I had met Veronica and Amalia on two prior visits to the office, but I had not yet been able to meet Don Alfonso, the new president of FERECOMINORPO, who had come to power just months before my arrival. He had recently been away traveling, and Amalia, a round woman with gray hair slicked back under her hat, had been promising that he would return soon. On this visit it seemed like he had finally proved her right. Veronica ushered me through the courtyard and up a set of stairs to knock on the building's rearmost door. "Pasa!" (Enter!) someone yelled from within.

Inside the office Don Alfonso was holding court in a seafoam-green office decked out with portraits of Simón Bolívar and Evo Morales, certificates from soccer championships, and a poster celebrating Llallagua's history that had been hand-decorated with glitter glue. A thin man with prominent cheekbones, Alfonso sat on the far side of a large desk, while several other members of the FERECOMINORPO directorate lounged in red plastic chairs along the room's edge. Alfonso did not look at me while I introduced myself and explained my interest in the history and politics of local mining cooperatives. Instead, he fiddled with a small desktop statue of a miner standing on a mountaintop. Tilted in one direction, the statue's arm fell to its side; in the other direction, it shot upward in revolutionary fervor.

Despite his seeming disinterest, Alfonso stood up when I finished talking. He walked around to the front of the desk and began to lecture in earnest. "*Mira*, if you're going to write about the history of the mining cooperatives, you first have to know the history of mining in the zone. And that means you have to write a story that starts in the colonial period" (Llallagua, January 25, 2016). Barely pausing to take a breath, Alfonso proceeded to narrate what I later came to think of as the origin story of Norte Potosí, which was repeated nearly every time someone discovered that I was interested in local history. In this case, Alfonso used his desktop statue as a teaching prop, jabbing the bobbing fist back and forth as he delivered a whirlwind tour through four historical periods: (1) (failed) colonial attempts to mine silver, (2) turn-of-the-twentieth-century private tin mining, (3) state-led tin mining following the 1952 National Revolution, and (4) cooperative mining in the aftermath of neoliberal state retreat. This last period was inaugurated with the 1985 passage of DS 21060, a number that Alfonso rattled off as automatically as his children's names. That supreme decree marked the dismantling of the state mining corporation COMIBOL and the layoff of more than twenty-three thousand unionized miners, who were, in Alfonso's words, *echados a la calle*—thrown into the street. Llallagua and Uncía, the town on the other side of the mountain, were both abandoned by the state corporation. In this postapocalyptic atmosphere, mining cooperatives formed in the shell of a mountain left behind by three previous waves of extraction. "Mining cooperatives emerged because they had to, because miners had no choice. We have always worked for our daily bread, and we always will," Alfonso concluded. Satisfied with his lecture, he leaned back and asked me if I had any questions, seeming to imply that I should have enough information now to wrap up my research.

I paused a beat, wondering how to explain that I had already read this history dozens of times but that it always struck me as descriptive rather than explanatory. I was in his office because I was looking for a different kind of story, one that attended to the administrative and collective efforts through which mining cooperatives had been constituted. I was especially eager to understand how mining cooperatives had become so internally stratified, given that they emerged from the ashes of famously collectivized workers' unions. There was no *necessary* reason why dismantling the unions should have resulted in cooperatives such as these, and my head buzzed with questions about what that moment of transition was like. Who had chosen the cooperative model, and why? How had miners negotiated subterranean access rights? Why had the mountain been carved up for multiple cooperatives, instead of one large one? Why were the boundaries between these cooperatives so fiercely defended today?

I was not able, in that moment, to articulate to Alfonso what I was looking for, at least not in words that coaxed him out of the origin story he was so accustomed to sharing. I left the office with permission to conduct research but without the sense that Alfonso really understood what I was up to. It took many months for a more complex origin story to emerge, one that did not dispense with Alfonso's historical skeleton so much as flesh it out. While his broad historical demarcations were useful, they neglected both the global contexts and the intricacies of local labor practices. National-level revolutions and laws remain important, but they did not emerge in a vacuum, nor were they applied to a blank canvas. Understanding how the local labor arrangement called *mining cooperatives* emerged in Llallagua requires more than national history; it involves tracing the historical dynamics of tin production and consumption across political scales and geological depths.

In this chapter I begin to explore the question of how tin-mining cooperatives operating in the same region—between the towns of Llallagua and Uncía, in the region of Norte Potosí—became so socially stratified. More precisely, I use historical evidence to understand why the seven mining cooperatives based in this region are hierarchically differentiated, with some cooperatives wealthier and more politically powerful than others (I save an examination of the cooperatives' internal differences for the subsequent two chapters). My wager is that this local hierarchy is best understood within a material history of tin, a metal that ties together local labor practices, national politics, and global economies. As developed in the introduction, a material history of tin is attentive to both tin's material qualities—the geophysical characteristics of tin ore and the chemical characteristics of the element Sn—and the material conditions under which the former (tin ore) was transformed into the latter (Sn, or a metallic commodity on the market). I am therefore drawing inspiration from both discussions of "resource materialities" and historical materialist debates about class composition and class consciousness.[1]

To yoke these two literatures together, this chapter develops the concept of the *material fix*, which explores the material traces left behind by repeated historical attempts to rearrange labor and technology to maintain the local economy amid international price fluctuations and declining resource availability. This concept builds on David Harvey's (2001) "spatial fix," which describes capitalism's drive to resolve internal crises through geographic expansion and restructuring. Playing with multiple meanings of the verb *to fix*, Harvey draws attention to the inherent contradiction between the mobility and fixity of capital, which needs to continuously expand yet relies on immobilized built environments for production and circulation. This contradiction can be temporally resolved (fixed) through new rounds of geographic restructuring, in which different regions are devalued and revalued to facilitate new rounds of accumulation.[2] Although the material fix also traces the geographic dimensions of capitalist processes, it differs from the spatial fix in three important ways. It is, first, a story told from a different perspective. If Harvey's spatial fix is written from the perspective of capital (i.e., its contradictions, drives, and resolutions), the material fix is written from the perspective of the place being fixed (i.e., local intentions and their physical manifestations). Second, this concept emphasizes the material remains of past fixes. The history of a place's repeated incorporation into capitalist relations shows up as a material palimpsest in three-dimensional space, where it continues to shape contemporary political arrangements. Third, the material fix is attentive to how efforts to resolve economic crises, regardless of whether they are implemented by capitalists or workers, can contribute to the emergence of new social hierarchies and new forms of exclusion. Capital is fixed not only in the built environment but also in the cracks drawn between distinct groups of workers, and both these fixes linger long after capital's departure.

Social stratification, of course, is far from unique to Bolivia's mining cooperatives. Rather, it is a recurring theme within ethnographic studies of artisanal and small-scale mining around the world. In Mongolia, for instance, religious leaders and gold traders have grown powerful performing spiritual and financial purification services for small-scale gold miners who must cleanse themselves and their money of misfortune accumulated while digging underground (High 2017). In Ghana complicated hierarchies of "shadow sovereigns" whose powers are anchored in the spiritual realm have emerged around the economy of *galamsey* (small-scale mining) (Coyle Rosen 2020). In a case particularly like the Bolivian mining cooperatives, small-scale gold miners in Sierra Leone are organized into hierarchical gangs at whose peaks are financiers and other nonminers (D'Angelo 2014). For the most part, ethnographic accounts like these balance an acknowledgment of the unevenly experienced social harms associated with small-scale mining against an analysis of the structural factors that make it an appealing—or even the only—livelihood option for many people. Building on insights from these ethnographies and others, this chapter is intended to provide a frame for exploring the historical emergence of labor hierarchies in a way that is not unduly critical of decisions made in contexts of increasingly limited options.[3]

My argument is that historical efforts to fix the failing tin mines of Llallagua-Uncía created a three-dimensional geography that continues to stratify local mining cooperatives today. Before exploring this material history, however, I begin by elaborating on the neoliberal origin story that Don Alfonso recounted in his office. In addition to allowing me to introduce the complications associated with studying mining cooperatives, this history is my entry point into a discussion of the structuralist interpretation of mining cooperatives that prevails in Bolivia. I proceed to complicate this origin story by showing how Llallagua-Uncía's three-dimensional geography was repeatedly transformed through supranational, national, and local attempts to fix economic and geophysical problems inherent to resource extraction—and how these historical fixes show up in the present.

NEOLIBERAL COOPERATIVES

In her study of the Argentine oil industry, anthropologist Elana Shever (2012, 11) opens with a discussion of the emic qualities of the word *neoliberalismo* in Latin America. In Argentina, as in Bolivia, *neoliberalismo* has become an everyday word whose syllables can be heard on a street corner just as frequently as in an academic lecture hall. It is used as shorthand to reference the privatization of state industries and services, the weakening of organized labor, and the removal of barriers to internal trade and foreign investment. But it has become such a commonsense category of analysis that *neoliberalismo* often seems like an agent in its own right, capable of tearing through social fabrics woven over decades, if not centuries. In the case of mining in Norte Potosí, *neoliberalismo* tore through an exceptionally resilient labor union fabric, with a warp of Indigenous community organizing and a weft of Marxist political principles (June Nash [1979] 1993). What lent *neoliberalismo* this power, and how were new mining formations cobbled together in the aftermath?

From a political economic perspective, early or "rollback" neoliberalism generally describes a set of economic precepts that privileged individual economic agency and free market exchange over state involvement in the economy (Peck and Tickell 2002). Unleashed globally in the 1980s, these edicts were initially targeted toward dismantling the institutions and safety nets built during the previous era of Keynesian economics and import substitution industrialization (ISI). While Keynesian economics was associated with the Global North, particularly the United States, ISI had been the economic foundation of rapidly industrializing countries throughout Africa, Asia, and especially Latin America—indeed, its principles were first outlined by Chilean and Argentine economists in the mid-twentieth century.[4] Both economic models involved the subsidization of domestic industries and expanded labor opportunities with major government-funded construction projects, but ISI placed more emphasis on holistic state economic planning and import regulation. With the global debt crisis of the 1980s—in turn spurred by the irresponsible recycling of petrodollars as loans to governments in the Global South, followed by a sudden hike in interest rates by the US Federal Reserve—ISI was no longer a viable option. The only financial relief for debt-stricken countries came with golden handcuffs: in exchange for their help, the World Bank and International Monetary Fund required strict economic structural adjustments of indebted countries (Gowan 1999). These conditions forced open national borders to free trade, drastically reduced state services, and privatized state industries. As state institutions and industries were shuttered or winnowed down to size, state employees had their salaries frozen or were laid off altogether.

Bolivia had the misfortune of being one of the first Latin American countries subjected to these economic "shocks," second only to Chile (Klein 2007). The policies unleashed were also some of the most extreme. Against a backdrop of collapsing tin prices, which had started to decline in the early 1980s, and hyperinflation in the wake of the debt crisis, New York economist Jeffrey Sachs designed a "stabilization" program that had dramatic effects on Bolivian society. Less than a month after having been sworn in as the country's first democratically elected president in two decades, Víctor Paz Estenssoro signed DS 21060 on August 29, 1985. Much more than a single policy, DS 21060 bundled together 220 separate laws referred to collectively as the New Economic Policy. It targeted hyperinflation by allowing the currency to float against the dollar, ended protectionist regimes, and privatized state-owned enterprises.[5] In one fell swoop, this decree seemed to undo three decades' worth of political organizing by workers' and peasants' unions across the country. Ironically, this was not the first time that Paz Estenssoro had been elected: he was also the first president to assume power after the 1952 National Revolution, the crowning moment of twentieth-century union activism. With DS 21060, Paz Estenssoro dismantled the very institutions he had helped to build.

The industry that was worst hit by DS 21060 was the tin-mining sector. Although the governance of other natural resources had never been strongly centralized in Bolivia, the tin-mining industry had been the country's economic pièce de résistance since its nationalization in 1952. Even when it was not as financially successful as had been hoped, it was still symbolically important, since its nationalization had been hard won following the combined efforts of workers' unions, peasants' unions, and middle-class activists. Today *neoliberalismo* in Bolivia is nearly synonymous with *relocalización*—a euphemism for the massive layoffs of unionized tin miners. It was called a relocalization because the miners were given severance packages that were supposed to help them begin their careers anew. Most miners, however, prefer to call it the *masacre blanca* (white massacre). Residents of Llallagua, in recalling this period, referenced the truckloads of people who began leaving on a daily basis, transforming the mining town into a ghost town nearly overnight. This was true across the highlands, as laid-off miners left in droves.

As they dispersed, ex-miners brought their knowledge of union organizing to other regions of Bolivia, a spatial repercussion of neoliberal economics that inadvertently helped lay the groundwork for "post-neoliberal" social movements. In the outskirts of major cities, ex-miners began organizing at the neighborhood or community level; they ultimately became integral players in Cochabamba's Water War (1999–2000) and the nationwide Gas War (2003).[6] Other miners traveled to the Chapare, a lowland valley outside of Cochabamba, where they began to cultivate coca. The new *cocalero* (coca growers') unions that they formed played an extremely important role in the political events of the early 2000s—Evo Morales himself was a *cocalero* union leader before he was elected president in 2005. Alongside Indigenous federations, the urban and rural collectives established by relocalized miners were essential leaders in the ousting of Gonzalo Sánchez de Lozada's neoliberal administration in 2003, which paved the way for Evo's election.

The celebrated stories of ex-unionized miners in anti-neoliberal struggles, however, do not generally mention that many laid-off miners went back to mining just a few short years after the passage of DS 21060. The ambiguous status of coca in the 1980s was a huge factor. Although coca is important in Quechua and Aymara communities, where it is used for quotidian, medicinal, and ceremonial purposes, this demand has been historically satisfied by the "traditional" coca cultivation zone, the Yungas valley north of La Paz (Spedding 2005). The rise of coca cultivation in the Chapare, on the other hand, is difficult to disentangle from the simultaneous escalation of demand for cocaine in the United States and beyond.[7]

The explosion of coca production after 1985 was a political problem for the Paz Estenssoro administration, which was relying on the goodwill of the international community to negotiate its way out of debt. The US-led war on drugs was already well underway across Latin America, but the evisceration of workers' unions increased the number of Bolivians willing to take the risks associated with coca cultivation. Anxious to lure ex-miners away from the coca fields, in 1987 the Bolivian government passed another decree (DS 21377) that permitted the formation of mining cooperatives in COMIBOL's abandoned properties. Miners flocked back to their former work sites. In Llallagua FERECOMINORPO—where I first met Don Alfonso—formed in June 1987 and today represents sixteen cooperatives with approximately four thousand members across the region. In Llallagua and Uncía alone, there are seven cooperatives (Siglo XX, 20 de Octubre, Dolores, Juan del Valle, Carmen, 23 de Marzo, and La Multiactiva) with a total membership of more than 2,600 miners.

Despite this neoliberal saga, however, mining cooperatives were not a new idea dreamed up in 1987—a point that is typically omitted from histories of mining cooperatives, including the origin story recounted by Don Alfonso. By the time that DS 21377 was announced, there were already more than three hundred mining cooperatives in Bolivia, and since 1968 the sector had been organized by a national federation known as FENCOMIN. As I demonstrate in subsequent sections, this pre-neoliberal history crucially informed the cooperatives' current organizational structure. But the moment of neoliberalization *was* transformative. As with the urban neighborhood associations and the *cocalero* unions, the absorption of so many recently laid-off workers had a major impact on FENCOMIN, which had previously kept a relatively low political profile. Miguel, a former union leader-turned-*cooperativista* from Llallagua, who became president of FENCOMIN in the early 2000s, described the moment of transition thus:

> When I entered FENCOMIN, I found everyone using neckties, it was obligatory to use one, but we ex-unionized miners don't like neckties much. So we needed to change the situation. Until I entered in 1998, FENCOMIN had never put forward a list of demands [*pliego*] like we did in the unionized camp. Before, some claims had been made, but they were quickly forgotten when the leadership changed. We [the relocalized miners] made demands in several new areas: about the obsolete machines and equipment that COMIBOL charged us for as if they were new, about unfulfilled promissory notes, environmental issues, insurance, and rent. And I promise you, all these things improved. (Llallagua, September 23, 2016)

Since cooperative miners are now notorious for their *pliegos* and their gruff, no-nonsense approach to politics—an attitude embodied by Miguel, who came to the interview with a stack of graphs proving that mining cooperatives are indispensable job creators and tax generators—it can be guessed that the swarm of ex-unionized miners indeed managed to change the institution's political culture. They also swelled its ranks: between 1986 and 1990, the number of mining cooperatives in Bolivia jumped from 325 to 586, with the new ones all located in the highland departments of La Paz, Oruro, and Potosí, in the former holdings of COMIBOL (Poveda Ávila 2014, 97). With more numbers and more organizing experience, FENCOMIN gradually became a major player in the national political scene.

Another ex-leader of FENCOMIN, Teodosio, narrated a story of gradual political awakening that gathered steam at the turn of the twenty-first century, when cooperative miners began to fight for incorporation into the social security national health care systems, for better machinery and equipment, and for assistance in prospecting new work sites. He elaborated, "[Mining cooperatives] were already a national reality, a generator of foreign exchange, not only of jobs; this is why we mobilized, to ensure that cooperatives were recognized as a fundamental pillar of the country's economy" (Llallagua, January 17, 2017). This process of obtaining official recognition culminated with the passage of the 2009 constitution, which declared mining cooperatives "productive actors" in the mining sector, on par with private and state corporations.

Almost exactly like the worker-owned *emprendimientos* (microenterprises) that Elana Shever (2012) describes emerging in the Argentine oil industry in the wake of neoliberal reform, cooperative miners embody a strange mix of familial care and entrepreneurial acumen that makes them difficult to locate politically. Unlike Bolivia's neighborhood associations and *cocalero* unions, mining cooperatives do not retain the same kind of solidarity that characterized the miners' unions. Instead of relatively egalitarian labor relations and redistributed profits—a common-sense expectation of workers' cooperatives—there are clear hierarchies between and within cooperatives. Each COMIBOL holding was divided among several cooperatives, and although each cooperative holds its concession as a collective, members do not pool supplies or profits or reinvest in production in any substantive way. Rather, membership—which can cost anywhere from US$50 to more than US$1,000, depending on the cooperative—gives miners access to the contracted area, after which they are free to move around and find their fortune underground.

In some regions of Bolivia, such as the city of Potosí, the cooperative labor structure has resulted in a kind of *tercerización* (outsourcing) whereby cooperative members accumulate enough capital to become stakeholders and employers rather than workers (Francescone and Díaz 2013). This structure is especially prevalent in gold cooperatives in the north of the department of La Paz, which emerged with the commodity boom of the early 2000s and grew astronomically with rising gold prices in the early 2020s, but traditional silver-mining cooperatives are also implicated in such practices.[8] For example, sociologist Neyer Nogales Vera (2009) has documented how some cooperative miners working in the Cerro Rico of Potosí were able to accumulate capital by laying claim to rich veins, investing in machinery, and contracting day laborers who worked for a percentage cut or a daily wage. In more extreme cases, the wealthy stakeholder did not work at all but rather managed his crew from afar, sometimes while pursuing a separate career in medicine, engineering, politics, or the like. More recently, economist Pablo Poveda Ávila (2014) published a review of forms of production across Bolivian mining cooperatives and showed that despite significant variation in labor organization, class differentiation within cooperatives is the norm rather than the exception. This differentiation is far from stable, a fact that is underscored by anthropologist Pascale Absi (2005) in her study of mining cooperatives in the Cerro Rico. She opens her book with the story of Don Fortunato, an aptly named miner who "won the lottery" by discovering a rich vein of silver—a finding that the other miners attribute to a deal with the devil. Don Fortunato quickly becomes a local celebrity who employs three hundred people and sponsors major town events, but by the end of the introduction, he is back working underground, his vein having dried up and his wealth having evaporated as quickly as it appeared.[9]

Regardless of its ephemerality, however, the image of a rich cooperative miner sitting comfortably in a two-story house while dozens of his hired laborers work in grueling conditions fuels much of the anti-cooperative mining sentiment in Bolivia. It would appear that salaried miners, having been granted control over their means of production following neoliberal restructuring, forgot their Marxist training and began to treat the mountain as their personal property rather than a shared national inheritance. How can such a dramatic shift be explained?

By and large, critics of mining cooperatives attribute the latter's reactionary politics to the same neoliberal policies that spawned them. This logic was crystallized for me in conversation with Eduardo, a professor in the FPS (political syndicalist formation) department of UNSXX (Universidad Nacional Siglo XX; National University "Siglo XX"), Llallagua's local university (founded by unionized miners in the 1980s). Eduardo was one of Llallagua's most outspoken Trotskyists, and every time I ran into him, he would ask me if I had finished reading Guillermo Lora's *Historia del movimiento obrero boliviano* (1980). Lora, born in Llallagua in 1922, was Bolivia's most ardent Trotskyist thinker and author of the *Tesis de Pulacayo* (Thesis of Pulacayo) ([1946] 2011), a landmark miners' manifesto that called for armed struggle led by the workers—in partnership with the peasants—to overthrow feudal landholders and the incompetent bourgeoisie. Lora, Eduardo promised me, was key to understanding the cooperatives.

On one occasion, frustrated by my lack of commitment to a *lorista* interpretation of the cooperatives, Eduardo took it on himself to give me a condensed lesson. We had bumped into each other on the street, and Eduardo used the wooden door next to us as an impromptu blackboard. With his finger, he drew two columns in the dust meant to represent mining cooperatives and miners' unions. "Cooperative miners," he said, pointing to the first column, "are like artisans: they do everything on their own. They mine the ore, they process the mineral, they take it to market. They each have their own *paraje* [work area]. When the work is too much for them, they hire others, acting more like small-scale capitalists than collectivized workers." Then, gesturing to the second column, he continued, "The unions, on the other hand, had a social division of labor. Some miners dug the ore, others processed the mineral, and still others took it to market. They had a completely different mode of production, and that makes a difference politically."

Moving to a different part of the wall, he drew an imaginary horizontal line. "Economic production has to be the basis of your argument. Then you can talk about the cooperatives' political organization" (a slightly higher horizontal line), "then about how they chew coca and worship the Tío and all that" (a tap above the two lines).[10] "But you need to begin with how they are landlords [*propietarios*] without socialized labor." Finally, he concluded his lesson with the exclamation, "They're landowners [*ter-*

ratenientes]! And their consciousness is an extension of their mode of production."

Eduardo's lesson condensed in simple terms (terms he likely thought appropriate for a gringa) a structural Marxism that is pervasive in the mines. In this interpretation the mode of production is figured as the base of society that determines political and cultural superstructures in the last instance. This simplification of historical materialism often draws on Karl Marx's writing in the preface to *A Contribution to the Critique of Political Economy* ([1859] 1904), a text that Marx himself abbreviates in a footnote of *Capital*: "My view is that each particular mode of production, and the relations of production corresponding to it at each given moment, in short 'the economic structure of society,' is 'the real foundation, on which arises a legal and political superstructure and to which correspond definite forms of social consciousness,' and that 'the mode of production of material life conditions the general process of social, political and intellectual life'" ([1867] 1990, 175, quoting himself).

It is notoriously dangerous to cite Marx since any given passage is best interpreted in the light of his entire oeuvre. In this case, the framework of structured levels presented above has been the subject of significant debate that, at its core, is about the content of class and particularly class consciousness. In their attempt to explain the absence of revolution, Marxists after Marx have explored at length the relationship between the objective conditions of class (also known by Marxists, if not Marx himself, as "class in itself") and the subjective consciousness of these conditions ("class for itself").[11] A narrowly structuralist Marxism might argue that this distinction is useless, because the objective conditions of class necessarily produce class consciousness; if consciousness has not been achieved, then the correct conditions must not yet exist. In other words, they might take the passage above at face value, arguing that political and ideological structures express the relations of more fundamental economic structures. By contrast, more cultural or praxis-oriented approaches to class consciousness have focused on the practices rather than the beliefs of struggle. Because workers rarely self-identify in exclusively class terms, Marxists in this latter group have emphasized the transformative potential of supposedly superstructural realms such as education, religion, gender, and racial and/or ethnic identity.[12]

Bolivian trade unionists and radical political economists, who tend to be informed by a Trotskyist reading of Marx, typically align more with a more structuralist interpretation of class consciousness.[13] That is, they distinguish carefully between class segments because they expect true class consciousness to arise within the waged working class, which ought to form a vanguard in any revolutionary struggle. This was explained to me by Javo Ferreira, a nationally prominent Bolivian Trotskyist whose book *Comunidad, indigenismo y marxismo* (Community, Indigenism, and Marxism, 2010) critiques Evo's government and rejects mainstream Indigenous politics by arguing that true decolonization depends on revolutionary class struggle. Between phone calls to members of the Trotskyist student organization, who were preparing to publicly contest Evo's bid to change the constitution to permit his reelection, Ferreira thoughtfully reflected:

> We Trotskyists have always insisted on the need to differentiate between classes. In the 1990s many sectors of society that were once part of the working class stopped being working class—in Bolivia [cooperative] miners are an example. . . . In many ways, the cooperatives are something like campesinos. When campesinos possess little land, they tend to get cozy with the working class, but when they have more land, they become more entrepreneurial. They are not reliable. Moreover, by blocking state development in the mining sector, they are functional to transnationals. They don't move with a national spirit, or a political consciousness. . . . When they're in precarious conditions, these nonworking classes often have political practices that resemble those of the left. But poverty does not make you working class! Right now, they are allied with the MAS [Evo's political party], but they are very pragmatic and individualistic. How long will their support continue? That's the subjectivity of poverty: say "yes" to the boss until you can plunge the knife. (La Paz, February 17, 2016)

Javo's analysis locates the rise of the precarious nonworking class immediately post–DS 21060 and identifies the risks of assuming that the ex–working classes will be as reliable as they used to be. By his analysis, mining cooperatives exemplify the "subjectivity of poverty," which seeks individual gains rather than collective entitlements. Thieves, Javo seemed to suggest, cannot have class consciousness.

While not always articulated so evocatively, this interpretation is widespread. It is even shared by some cooperative miners, particularly those who remember what it was like to be in a union. For instance, when I asked Teodosio, whom I quoted earlier, how ex-unionized miners could have become so individualistic in their mining practices, he shook his head and responded:

> I ask myself the same question. If we of Siglo XX [a former miners' union, now a cooperative] once reached the maximum expression of unionism, where did the solidarity go? Individualism beat us. From the collective conception, we regressed in our degree of organization. He who had the vein did not share with his companions, though he worked with family and close friends. The problem of individual production can be overcome with collective production, but that's one of the great weaknesses of cooperatives. . . . The Supreme Decree 21060 was the root of individualistic thinking, the law of supply and demand, and breaking with it would be a very long process. (Llallagua, January 17, 2017)

While the impact of neoliberalizing policies on Bolivia's mining sector can hardly be denied, this causal narrative—that the policies changed the economic structure, which changed miners' labor organization and their consciousness—is too simple. It does not explain why neoliberal policies were so geared toward the tin mines or why mining cooperatives took the form they did. Even more, it stifles a nuanced engagement with the practices of cooperative miners by assigning all-encompassing agency to a set of policies enacted over thirty years ago. As the quote above shows, it can even suppress political transformation within cooperatives, whose members see their fate as historically sealed.

In the following sections, I excavate other histories, beginning—as Don Alfonso suggested—much earlier than 1985. But rather than just starting in an earlier period, I begin with a different agent: tin, the metal whose multiscalar dynamics shaped internal class dynamics within both the unions and the cooperatives.

TWENTIETH-CENTURY FIXES

When Bolivian prospector Simón I. Patiño discovered a massive vein of black ore in his tiny mining concession in the Juan del Valle Mountain at the turn of the twentieth century, he apocryphally whispered to himself while waiting for the laboratory results: "*Que no sea plata, que no sea plata* [Don't let it be silver]. Dear God, let it be tin" (Querejazú Calvo [1977] 1998, 45). Although silver mining had dominated Bolivia's economy since the start of the colonial era, times were changing. Over the next several decades, the Juan del Valle Mountain's geological history was articulated with new developments in global political economy, transforming the mountain into a mine and generating a workforce whose echoes can be found in today's mining cooperatives. The particular qualities of the ore in the mountain's folds made it valuable in the global marketplace and called forth a set of workers who were shaped in the mountain, the crucible where local geology was articulated with global capital.

By the time Patiño was praying for tin, it was far from a novel metal. Tin had been used to stabilize metal alloys in Europe and the Middle East since ancient times and was even known in Latin as *diabolus metallorum* (the devil's metal) because it could make any alloy brittle (Duffield et al. 1989, 147).[14] When it is mixed with copper, tin creates bronze, a durable metal that became so ubiquitous that it is now used to define the historical era from around 3000 BCE to 1200 BCE. Yet tin mines were not objects of desire like gold or silver, in part because a steady stream of tin from Cornwall satisfied most demand (Baker 2018; Ingulstad et al. 2015). This began to change in the mid-eighteenth century, when tin became an essential ingredient for industrial capitalism. No longer just a useful alloying agent, tin became useful as a dye fixative in the textile industry, in solder and bearing metal used for electrical and mechanical construction, and in a variety of military attachments (Ross 2014). Above all, however, tin was used to preserve food. Unlike most metals, tin does not oxidize when exposed to air or water, and it was therefore used to package food for urbanizing populations, frontier colonists, and the armed forces. With these drivers, global tin production rose from 36,000 to 124,000 tons between 1874 and 1914 (Duffield et al. 1989, 147). After two thousand years of continuous production, the Cornish tin mines coughed and sputtered into exhaustion, creating a tin shortage that prospectors worldwide were eager to fill.

This was the context in which Patiño struck tin in the Juan del Valle Mountain. According to the stories that miners tell about him, the vein he exposed was several meters wide and pure enough not to require preliminary processing; it could be shoveled directly into sacks and transported, via llama, to Oruro. Patiño named the vein La Salvadora (The savior) because it saved him from certain penury, as he had spent all his wife's inherited fortune prospecting without luck. Albina Rodríguez, a wealthy member of the Oruro elite, had sold her family's jewelry to support him (Querejazú Calvo [1977] 1998). But their patience was richly rewarded:

producing ore that contained more than 50 percent tin, La Salvadora was one of the purest tin veins in the region, if not the world (Geddes 1972, 63).

Over the next twenty years, Patiño expanded his operations by buying out the other smaller companies around him. On July 5, 1924, he became the sovereign regional tin magnate with the incorporation of Patiño Mines Enterprises in Delaware, a move that also transformed him into one of three Bolivian oligarchs known collectively as *los barones del estaño* (the tin barons) (Mitre 1993). By the late 1920s, Patiño and the other two tin barons, Moritz (Mauricio) Hochschild and Carlos Víctor Aramayo, had propelled Bolivia to the top of global tin production, vying only with the Dutch East Indies (Indonesia) and the Federated Malay States (Malaysia). The Patiño holdings supplied 60 percent of Bolivian tin, of which three-quarters came from the Juan del Valle Mountain (Mantell 1949, 80). To celebrate its modernizing capacity, Patiño named the Juan del Valle mine Siglo XX (Twentieth Century). Now that we are well into the twenty-first century, this name sounds strangely anachronistic, yet in Llallagua it is still attached to the largest mining cooperative, a major mine shaft, the local university, and the section of town that was once worker housing.

News of Patiño's lucky strike spread quickly, and workers came traveling on foot and llama from Cochabamba, Oruro, and Potosí. Many came with their families, supporting children, spouses, and aging parents, seeking jobs maintaining machinery, building houses, drawing maps, and working underground. Llallagua and Uncía emerged on either side of the Juan del Valle Mountain, two company towns that more properly resembled archipelagos of production areas, worker housing, administrative offices, company stores, and other necessities. In one section of Llallagua, known as Catavi, management offices were flanked by a theater, a golf course, a top-tier hospital, and elegant bathhouses built on natural hot springs, all for the exclusive use of managers, engineers, and their families. In another section, known as Siglo XX, miners and their families shared one-room adobe houses with limited access to fresh water, latrines, and heat. In this way, Patiño not only fixed his workforce in place but also ensured that its economic echelons were materially etched in the landscape.

Although Patiño hired primarily men to work in his mine, he also hired a fair number of women, particularly the teenage daughters of mining families. These women worked as *palliris*, which means that they sorted through rocks brought to the surface to determine which contained valuable ore and which ought to be sent to the slag heaps. *Palliri* is a colonial labor category that derives its name from the Quechua word *pállay*, which means to harvest or to pick up from the ground, regularly Latinized as *pallar*.[15] These women were crucial to Patiño's enterprise, since they limited the amount of tin lost to the slag heaps. As contributors to their family's income, they also helped maintain households that would not have been able to survive on one wage alone.

But a workforce was only part of what Patiño needed to make his enclave economy operational. He also needed a means of commodity circulation. The Bolivian government was in the process of building railroads to connect the major centers of production with urban areas, but construction was not happening quickly enough for Patiño. In 1914 he began building his own railroad line to connect Uncía with Machacamarca, a nearby town on an existing railroad line. This railroad was a crucial means by which he exported ore and imported machinery, building supplies, and new workers. Investments in physical infrastructure increased profits; they also drew even more workers to the region. Geographically distant from urban centers and seats of government, Patiño's mine was nonetheless the infrastructural and economic heart of the nation, and it was thriving.

Or at least it thrived for a while. Although tin prices soared during World War I, they slumped in the immediate postwar years. In 1931, in an attempt to control the unpredictable mineral market, the first global tin cartel was established with four founding members: the Federated Malay States, Nigeria, the Dutch East Indies, and Bolivia. Like the more famous Organization of the Petroleum Exporting Countries (OPEC), the International Tin Committee (ITC) was made possible by the geographic and geological qualities of the resource it sought to control. Tin deposits of commercially viable quality were present in only a few countries, and internal demand for tin in these countries was low, meaning that producers' and consumers' interests were easily separated along national lines. Moreover, the market crash of 1929 had driven small tin producers out of business, and relatively stable tin oligarchies had emerged in nearly every country (except Malaysia). As the wealthiest of Bolivia's three tin barons, Patiño became his nation's unelected spokesman within the ITC. Over the course of the next decade, ITC representatives met three times (in 1933, 1938, and 1942) to update and confirm their shared commitments. In the same period, they were joined by three new member countries (Thailand, Belgian Congo, and French Indochina). Production capacities were renegotiated at each of the three meetings, and a buffer stock was maintained with increasingly formalized control (Mantell 1949). Altogether, this was a supranational fix through which the cartel attempted to retain the value of their resources against the hazard of a globally volatile market.

Patiño seemed to have fixed his resource, his workers, and his market, but by the 1940s all of these were proving difficult to control. The fixes themselves were causing problems. First, the more tin he mined, the lower the quality of the fresh ore extracted from the mountain. When Patiño began excavating La Salvadora in 1901, he and his workers simply followed the veins, but this technique only worked while the veins were thick and rich. By the 1940s Patiño had introduced two technological changes to deal with declining ore grade: a new method of extraction called *block caving* and a new method of ore processing called *sink and float*. Block caving involved imposing a three-dimensional grid on the whole mountain and detonating cubes of subterranean mountain. When it went smoothly, the broken rocks tumbled down through carefully carved chutes and into trolleys that carried both the ore and overburden (waste rock) to the surface. When it did not go smoothly, large rocks blocked the chute and had to be unjammed by miners with long metal poles, an extremely dangerous job that resulted in many deaths. And even when it did go well, block caving also created a new problem: How could low-grade tin be separated from so much overburden? To facilitate processing, in 1945 Patiño built the Sink and Float Plant, which relied on the chemical xanthate to separate tin, which sank, from waste rock, which floated. This plant was disastrous for the *palliris*, since their expertise in distinguishing tin ore from waste rock was no longer as valuable: even rocks that looked like waste often contained enough ore to justify chemical processing. *Palliris* continued to roam the growing *desmontes* (slag heaps) looking for tin, but their work was increasingly marginalized within the sprawling company operations.

Second, Patiño's employees began to organize. Via his new railroad, Patiño had unintentionally been importing political theory alongside dynamite and building supplies. The trains that came to Uncía were operated primarily by unionized Chilean and Argentine workers who were eager to share their knowledge of Marxist and Trotskyist political theory with the Bolivian tin miners. Within a remarkably short period of time, the miners established what would become two of the world's most powerful workers' unions (Smale 2010): Siglo XX, composed primarily of underground workers, and Catavi, composed primarily of workers from the concentration and processing plants. When miners returned from the disastrous Chaco War (1932–35), the unions were strengthened further still. In 1944 miners from across the country met at the Huanuni tin mine—another Patiño holding, located down the road from Llallagua—and created the FSTMB (Federación Sindical de Trabajadores Mineros de Bolivia; Syndical Federation of Mine Workers of Bolivia). Juan Lechín, a miner from Llallagua, was chosen as the first leader of the FSTMB, since the Llallagua unions were the most militant of them all.

Third, the ITC was losing its international grip. Threatened by the cartelization of tin, the United States had developed a plan for stockpiling the newly precious metal. In 1939, with the passage of the Strategic and Critical Materials Stockpiling Act, the United States began to build up reserves of metals that had proved scarce in World War I. Bolivia's ore was of particular interest, given its relative proximity to the US sphere of influence. With the outbreak of World War II, Japan took control of Malaysian tin mines at the same time that Atlantic shipping routes became too vulnerable for Bolivians to continue shipping to smelters in Europe. By 1942 Bolivia was the only supplier of significance to the Allies and was forced to ship its product to smelters in Texas, where prices were kept artificially low. By the end of the war, substantial Bolivian deposits had been mined at unfavorable prices (Dunkerley 1984, 11–12). Meanwhile, the United States had stockpiled 350,000 tons of the metal, equivalent to world consumption for two years (Mallory 1990, 839).

All this set the context for the National Revolution of 1952, one of the crowning achievements of which was the nationalization of Patiño's mines and the creation of the state mining corporation, COMIBOL. But the mine that COMIBOL inherited in Llallagua-Uncía had already been materially shaped by Patiño's fixes—and their limits.

POSTREVOLUTIONARY FIXES

The National Revolution is typically the background of the mining cooperatives' neoliberal origin story, with the unionized miners acting as the revolutionary foil for the cooperatives' individualized thieves. But on closer inspection, cracks emerge in the story of worker unification in the postrevolutionary decades. These cracks both provoked and were provoked by multiple fixing efforts that slowly deformed the mountain and its workforce.

The financial troubles of COMIBOL started almost as soon as the post-revolutionary dust had settled in 1952. Still headed by Juan Lechín, the FSTMB had won "workers' control" over the new state mining corporation, which meant that FSTMB was involved in decision-making processes and even had veto powers. One of the FSTMB's core demands was that COMIBOL hire back all the miners who had been fired for political or medical reasons. The salaried workforce swelled from 24,000 in 1952 to 35,660 in 1956 (Espinoza Morales 2013, 56). In a context of falling tin prices, this threw the country's economy into disarray. As Bolivian union leaders reminded me, the financial situation was not the fault of the workers: they had demanded that the tin mines be expropriated without compensation, but the MNR had chosen the more internationally palatable option of a forced sale.[16] In its early years, COMIBOL had to keep paying former mine owners in annual installments that were a huge blow to profit margins (Canelas Orellana 1981, 60–61).

These internal problems were exacerbated by changes in the international market. After World War II, prices fell as trade embargos lifted and military demands diminished. The recycling campaigns that European and North American countries had been waging domestically during the war also affected prices, since consumers continued these practices long after peace was declared. But the real crisis for tin came in the form of commodity substitutes. The biggest threat was aluminum, found in bauxite ore. Aluminum is the most abundant metal on earth, but the chemical and electrolytic processes necessary to separate the element from the compound were not established until the late 1880s and were not widespread until the mid-twentieth century. By 1965, however, aluminum-based "tin-free steel" was starting to replace tinplate steel throughout the canning sector, and demand for tin fell dramatically.

In the face of COMIBOL's mounting debts, the new Bolivian president, Víctor Paz Estenssoro, approached the international community for financial assistance. This was the same man who would later sign off on DS 21060, but in the mid-twentieth century, he was still trying to balance a center-left political program with a constrained financial reality. In 1961 his diplomatic efforts crystallized in the Triangular Plan, a US$62 million program that was eventually subsumed into the US-led Alliance for Progress. Conceivably a precursor to DS 21060, the Triangular Plan aimed to "rehabilitate" the flagging COMIBOL through a series of massive worker layoffs. Melvin Burke (1987) highlights that the unions were initially amenable to the layoffs, since it was clear that some workforce reduction was needed, and the compensation packages were generous. Their attitudes changed, however, as the plan progressed and the government's methods grew violent.

One of the problems the Bolivian government faced as it cut its workforce was that laid-off workers would not go away. Instead, they kept reappearing in the mines as *jukus*, a Quechua word that translates as "birds that fly by night" and implies theft (Kohl, Farthing, and Muruchi 2011, 36). *Jukeo* was not a new phenomenon and had in fact been a largely accepted practice since Patiño's era. *Jukus* expected nighttime access to the mines as a moral economic right and actually lined up in the evening to enter as soon as the guards were off duty (Godoy 1985a).[17] Since *jukus* were often teenagers and recent migrants, one can imagine that *jukeo* initially worked in the company's favor, since these supposed thieves required no training or wages and generally sold the ore back to the company from which it had been stolen. Nevertheless, the growth of *jukeo* in the 1960s aggravated the already existing financial troubles of COMIBOL. The government wanted to cut more jobs—although five thousand miners had been laid off by 1965, COMIBOL was still in the red—but it also wanted to stem the growth of *jukeo*.

In this context, General René Barrientos Ortuño, who had come to power in a 1964 coup, launched a full-scale assault on the tin miners. He began by requesting a report on the state of Bolivia's mining sector from Roberto Arce, a mining engineer who had worked as a manager at Patiño Mines prior to the nationalization. Arce's (1965) scathing review of COMIBOL recommended that worker unemployment generated by the Triangular Plan should be dealt with by creating "autonomous societies" that could mine independently from the state corporation, absorbing laborers and relieving the corporation of the burden of paying salaries. The unions waged a vicious fight against this proposed solution to COMIBOL's financial crisis, with the most violent confrontation occurring in Llallagua on June 23, 1967. Soldiers were dispatched during the nighttime festival of San Juan, typically one of the coldest nights of the year, and shot miners and their families amid the celebrations. Official records report around twenty deaths and seventy injuries, but Llallagueños (people from Llallagua) I spoke with suggested that at least a hundred people died. Although this massacre is typically linked to the search for Che Guevara, the Marxist guerrilla leader who had been at large in Bolivia since 1966 and to whom the tin miners were considering offering their support, it was equally aimed at weakening the unions that continued to oppose the Triangular Plan.

Despite the unions' resistance, COMIBOL took Arce's advice. Instead of calling the collectives of independent workers "autonomous societies," however, COMIBOL called them either *subsidiary organizations* or *mining cooperatives*. When the Triangular Program ended in 1970, COMIBOL had forty-eight cooperatives on its property, which collectively had five thousand official members and four thousand associate members (Burke 1987, 33). Meanwhile, FENCOMIN was founded in 1968, representing eighty mining cooperatives with a total of twenty thousand members (FENCOMIN 2008). By not only legalizing but actively institutionalizing mining cooperatives, COMIBOL reduced its expenses and neutralized the unions. The cooperatives thus emerged first as a material fix adopted by the Bolivian government to preserve the profitability of its failing state corporation and ensure continued US aid money in the postrevolutionary period.

At the same time that COMIBOL was creating cooperatives from scratch, it also began formalizing preexisting groups of independent and small-scale miners who had always worked on the margins of mining properties. Indeed, as cooperative miners often remind me, the oldest cooperative in the country was founded in the city of Potosí in the immediate aftermath of the global economic crash of 1929 (Mariobo Morena 2007). They rarely note, however, that this group was not technically a cooperative when it first formed; rather, it was a "union of free workers" claiming a lineage with *k'ajcheo*, a practice that began as organized ore theft by miners in the sixteenth century and slowly evolved into a customary right in the seventeenth century.[18] As with *jukeo* in twentieth-century tin mines, *k'ajcheo* was customarily accepted in colonial and early republican silver mines, and mine owners factored it into their financial calculations. By the economic crisis of the 1930s, however, many unemployed miners had turned to *k'ajcheo* as a full-time livelihood strategy, and they formed the collective K'ajchas Libres y Palliris (Free K'ajchas and Palliris) to combine their extractive efforts and avoid the commitment of selling their ore to a single processing plant. They did not become a cooperative until 1958, when the Bolivian government passed the General Law of Cooperative Societies (Law 5053), after which K'ajchas Libres y Palliris functioned as something of a template for the many autonomous societies and cooperatives that began in earnest in the 1960s. The institutionalization of *k'ajcheo* transformed it from a colonial labor category that expressed disruptive worker agency into an instrument designed to preserve the functionality of dominant systems of state-led and private mining.[19]

This history is not unique to *k'ajcheo* or to Bolivia. For instance, Robyn d'Avignon (2018) describes how, in colonial West Africa, the practice of *orpaillage*—which is translated as "gold panning" but is only used to describe the practices of African miners, even when Europeans used the same tools—was encouraged within low-yield areas at moments when it benefited colonial mining interests. Similarly, Jody Emel, Matthew Huber, and Madoshi Makene (2011) trace how moments of contracted state capacity in Tanzania have corresponded with benevolent policies toward artisanal and small-scale mining. In both these contexts, unemployed people and low-value resources were paired to stabilize dominant industries through periods of relative recession. These old material fixes continue to reverberate in the present, as small-scale miners tend to persist in their labors long after they have lost official state support.

In Llallagua-Uncía the autonomous societies that began to emerge in the 1960s were called *subsidiary organizations*. When economic anthropologist Ricardo Godoy (1985b) conducted fieldwork in the region in the early 1980s, members of these subsidiary organizations were so numerous that he described the unions as a comparatively small regional labor aristocracy. By allowing the company to lease the most exhausted or least productive areas of the mine to independent workers, subsidiary organizations "fixed" the twin problems of ore degradation and excess workers. Subsidiary organizations sold all their tin ore back to COMIBOL, which deducted a percentage as rent; for the company, this was cheaper than paying wages. But subsidiary organizations were also welcomed by some workers, particularly those who had been too young, too old, too seasonally unavailable (because of agricultural labor), or too female to work as salaried miners. Indeed, most of the subsidiary organizations were created to meet workers' demands as well as to solve the company's problems.

In order of their historical appearance, there were three groups of subsidiary workers: *veneristas*, *lameros*, and *locatarios* (Godoy 1985b; Harris and Albó 1976).[20] Their organizations were often called unions, but only out of habit; the workers were not *regulares*, or waged company workers. *Veneristas* worked in the ore sands left in a dry riverbed around the base of the mine, *lameros* worked in the wastewater ejected by the Sink and

Float Plant, and *locatarios* worked underground in less profitable areas of the mine, but all three groups shaped the mining cooperatives that emerged in the late 1980s.

Veneristas worked in the liminal space of ore sands at the base of the mountain, digging holes that were up to fifty meters deep straight into the ground. This was a dangerous job: although the holes were bolstered from within with wooden beams, accidents frequently occurred when these beams collapsed, or when water levels suddenly rose. The oldest mining cooperative in Norte Potosí (and the second oldest in the country after K'ajchas Libres y Palliris, according to locals) is based in Uncía and traces its history back to a group of *veneristas*. Called simply Juan del Valle, this group of miners was registered as a cooperative in 1961, shortly after the passage of the General Law of Cooperative Societies and immediately prior to the Triangular Plan (Poveda Ávila 2014, 29). However, one of the cooperative's founding members explained to me that this was just part of their story: what became a cooperative in 1961 had existed as an informal group of ex-company miners known as Centenario since at least 1950. At its height, around the time of the National Revolution, it had 2,500 members, or roughly half of the population of Uncía (Uncía, February 19, 2016). In other words, this cooperative predated the Triangular Plan by a decade, though it was given new institutional life with the plan.[21] It had been "fixing" the problem of superfluous employees since before the National Revolution, and its role expanded when it received formal recognition.

Lameros were institutionalized as the *veneristas'* riverine counterparts in 1963. Like *veneristas*, *lameros* stayed close to the surface of the mine, though *lameros'* labors consisted in capturing tin from the wastewater that flowed out of the Sink and Float processing plant rather than digging holes in the ore sands. Both organizations, moreover, were initially conceived outside of COMIBOL and only subsequently recognized by the company. Hugo, a retired miner who began working as a *lamero* when he was fifteen, explained to me that the leader of the Siglo XX union, Federico Escobar, had proposed that COMIBOL institutionalize an organization of *lameros* to assist with rising regional unemployment (Llallagua, July 5, 2016). With the company's approval, members of the new subsidiary organization divided the river into segments that were each between two and three meters long. Every morning, the *lameros* lined up next to their assigned segments and, at nine o'clock sharp, jumped into their respective segments and began shoveling out *lamas*, the slimy silt that still contains small amounts of tin settled on the river floor. Since the segments of the river nearest to Sink and Float Plant had the richest mud, *lameros* rotated segments each week, with those who had been in the back going straight to the front. The institutionalization of this work benefited many people, particularly women and teenagers who were ineligible to work underground, and Hugo teared up recalling Escobar's contribution to the people's cause. But allowing *lameros* to work in its wastewater also benefited the company, which managed to capture tin that would otherwise have been lost downstream while also incorporating new, unsalaried laborers into the corporate complex.

Finally, *locatarios* (literally "tenants") were officially organized near the end of the Triangular Plan. I learned about their history from Andrés, cofounder and ex-leader of the *locatarios*, who recounted it to me over tea in the lobby of the hotel where I rented a long-term room. Andrés explained that during the Triangular Plan, General Barrientos (who had seized power from Paz Estenssoro in a coup in 1964) had gifted sections of the mountain to thirty-one individual (often foreign) prospectors known as *arrendatarios* (another word for "tenants"). These *arrendatarios* proceeded to hire miners who had been laid off from the state corporation to work in their private concessions. Persecuted for his work in the labor union, Andrés ended up working for an *arrendatario*, an Italian who was operating the Dolores section in the upper reaches of the mountain, where ore is of relatively poor grade. Andrés was angry that after all the miners' revolutionary efforts, they were once again working for private prospectors. On October 20, 1969, he and his fellow workers staged a takeover of the mine and expelled the *arrendatarios*, declaring themselves owners of the underground space. This coincided with a moment of dramatic political upheaval nationally: Barrientos had died in a plane crash in April 1969, and after three successive military coups, left-leaning General Juan José Torres had come to power on October 6. Instead of reinstating the *arrendatarios*, Torres transformed the occupying collective of workers into *locatarios*; they called themselves 20 de Octubre to commemorate their initial occupation (Llallagua, June 12, 2016). As with *lameros* and *veneristas*, it was workers' activism that created space for *locatarios*, but in many ways it was the state corporation that benefited, since *locatarios* reduced the company's cost of exploitation in low-grade areas. Arguably, this could explain the success of the *locatarios*' uprising: the corporation did not have anything to lose in low-quality sections that had already been gifted to private *arrendatarios*.

These subsidiary organizations helped make COMIBOL profitable again, but—as with previous rounds of fixing—this was a temporary solution. At the beginning of the 1980s, the United States began selling its stockpile of tin and dramatically depressed global prices by dumping more than thirteen thousand tons on the market. The ITC tried to buy all the tin to stabilize prices, but this involved extensive borrowing from commercial banks. The ITC hit its credit limit in 1985, triggering a massive crisis for tin: prices collapsed so quickly that tin was delisted from the London Stock Exchange for about three years (US Geological Survey 2013, 181). This constituted a market disruption that neither COMIBOL nor the Bolivian state could weather with subsidiary organizations alone. It paved the way for neoliberalization by disrupting the Bolivian economy sufficiently to make DS 21060 appear necessary.

The specific form that neoliberalization took in Llallagua-Uncía, however, had been shaped by a much longer history. Layered on top of Patiño's early twentieth-century fixes, the Cold War–era fixes of the 1960s–1970s created the sociospatial architecture within which contemporary mining cooperatives emerged. As I show below, all these fixes remained materialized in the earth itself.

NEOLIBERALIZATION: A PALIMPSEST OF MATERIAL FIXES

The Juan del Valle Mountain was geologically uneven long before miners began to slowly exhaust its tin reserves, and this uneven patterning has shaped all subsequent attempts to fix it. A Llallagueño geologist, Fredy, explained to me that the presence of tin ores in the mountain is associated with its mineralized "stock," which describes rock that was formed as an igneous intrusive—that is, molten rock that was shot up into preexisting sedimentary rock and cooled underground. It is a bubble of magma inside a shell of ancient lake bottoms. The stock, Fredy went on, is like the *criadero* (nursery) for tin: inside the stock, metalliferous veins grow like tree branches. He drew a sketch that looked like an upright pear to show what the stock would look like if it were possible to see it in vertical cross-section, and drew lines extending across the pear to illustrate how hot water and gases prompted the formation of metalliferous veins within the stock. These veins, Fredy emphasized, originated within the stock but extended well beyond it, into the surrounding sedimentary rock. In the sedimentary rock, however, the vein's mineral content was lower; the vein might appear equally thick, but its market value was greatly reduced.

What Fredy was sketching was three-dimensional variation in sub-terranean mineralization. Tin does not occur in a "native" form to any extent—a geological way of saying that tin cannot be found as pure crystals—but it *does* occur in a wide variety of ore types, each of which has a different concentration of tin (Mantell 1949). The closest approximation to pure tin is the ore cassiterite, also known as stannic oxide (SnO_2). Cassiterite has only to be shaken loose from its surrounding rock before it can be sent to the smelter; this is the ore that Patiño discovered in 1901, and this is the ore that is found in the metalliferous veins within the stock. At a much lower concentration, however, tin can also be found in sulfides such as stannite, which miners call tin pyrite. In addition to having less tin per ounce than cassiterite does, stannite also has to be processed chemically before it can be smelted.

This seemingly insignificant geological distinction has had a huge impact on the history of the mine and the class formation of today's cooperative miners. In the Juan del Valle Mountain, the stock is thicker at the deeper levels than toward the surface, which means that there is significantly more cassiterite in the depths of the mountain than in its higher reaches. Like most mines that are excavated from the inside (rather than through the formation of an open pit), this mountain has been divided into horizontal slices, each about thirty meters high, that start from the peak (level 0). These slices are representational guides in diagrams of the mountain's interior space, but they are also materially demarcated by mine shafts, elevators, and other subterranean thoroughfares. The lowest workable level is eight hundred meters down from the peak (level 800)—below that, there is too much water, even with multiple pumps. The richest workable levels are between levels 650 and 800.

When the subsidiary organization of *locatarios* was created in 1969, they were granted the upper levels of the mountain, while the lower levels were maintained by the company and worked by salaried employees organized in the union known as Siglo XX. With the neoliberal gutting of COMIBOL in 1985, the organization of *locatarios* was transformed into two cooperatives, one of which retained the *locatarios'* name—20 de Octubre—while the other took the name of the mine section, Dolores. Same people, same labor structure, same deposits: for ex-*locatarios*, the absence of the company changed only their market practices, since they had to begin selling ore to independent commercializers rather than COMIBOL. These two upper-level cooperatives, however, were joined by a third, Juan del Valle, which was composed of ex-*veneristas*. The *veneristas* had long exhausted the mineralized deposits of the former riverbed, and DS 21060 worked in their favor since they were able to secure access to richer subterranean deposits. Meanwhile, the ex-unionized miners formed their own cooperative that kept the name Siglo XX and continued to operate where they had been working before, in the lower levels of the mountain.

The mountain's subterranean space was thus divided among four cooperatives: three upper-level cooperatives composed of former subsidiary workers and one lower-level cooperative composed of formerly unionized miners. What had been a division between waged miners and subsidiary workers was solidified as a spatial separation underground, and the hierarchy between them was maintained in differentiated ore quality. Since the tin deposits are richer on the lower levels of the mountain, *cooperativistas* working in the higher levels have lower-quality tin at their disposal. These miners therefore produce less value with the same amount of labor, given that they must extract more ore to get the same amount of tin. But there is more to this inequality than unequal profits. Ore from the lower levels is purer and can be processed relatively quickly using just water and gravity, whereas ore from the higher levels contains sulfur and must be processed with gasoline, hydrochloric acid, and xanthate. Individuals working in the higher levels have hands rubbed raw from chemical exposure, and they claim to have higher rates of respiratory problems from inhaling the xanthate fumes. The inequalities generated by Cold War–era subsidiary organizations are preserved in the mountain's three-dimensional space and in the bodies of cooperative miners, who remain divided by wealth and by toxic exposure.

A similar process played out on the surface. Prior to 1985, the surface was worked by *veneristas*, *lameros*, *palliris*, and the Catavi union (salaried workers in the processing plants). After 1985 the *veneristas* moved underground, as already noted. *Lameros*, who had lost their tin reserve because there was no longer a steady stream of wastewater exiting the Sink and Float Plant, combined forces with the *palliris*, who had already picked most of the valuable rocks from the slag heaps. Together, they began operating artisanal concentration systems that extracted tin particles from the ore sands piled on the edges of town. They formed two mining cooperatives, Carmen and 23 de Marzo, both of which integrate former *lameros* and *palliris* and occupy the region that was once a floodplain of wastewater. Meanwhile, the Catavi union formed La Multiactiva, a mining cooperative that holds and operates the Planta Victoria, a concentration plant where they process wasted ore sands through machines once

TABLE 2.1. Transitions from Pre-1985 Mining Collectives
to Post-1987 Mining Cooperatives

Pre-1985 Mining Collectives	Post-1987 Mining Cooperatives
Siglo XX union (salaried)	Siglo XX
Catavi union (salaried)	La Multiactiva
20 de Octubre union (*locatarios*)	20 de Octubre Dolores
Canal workers' union (*lameros* and *palliris*)	Carmen 23 de Marzo
Juan del Valle (*veneristas*)	Juan del Valle

Source: Prepared by author

owned by COMIBOL. La Multiactiva, although small compared to Siglo XX, is widely considered the wealthiest cooperative in Llallagua-Uncía, and this comparative wealth is due in no small part to its members' history of salaried, unionized labor—and their ongoing control over their former work sites.

In this way, historical divisions between salaried and subsidiary workers that date back to the 1950s have been transformed into differences between mining cooperatives, and these differences are (re)produced spatially. The material fixes introduced to resolve the crises of inadequate ore, falling prices, and excess labor also created the material scaffolding within which mining cooperatives developed. The geologically varied depths of the mountain, as well as the surface of the mine, have shaped and been shaped by a century's worth of extraction; today's mining cooperatives crystallize much older socionatural disparities (see table 2.1 for a summary).

NATURE, SPACE, AND TECHNOLOGY

This chapter traced a history of tin mining in Llallagua that focused on how crises in the tin-mining sector were managed with technological innovations, local infrastructure, supranational cartels, worker layoffs, and subsidiary organizations. Each material fix addressed not only the contradictions of capital (and the corresponding fluctuations in the metal market) but also the specific challenges of extracting a dwindling, nonrenewable resource. If the premise of David Harvey's spatial fix is that capitalism has a spatial contradiction—it requires that material infrastructure be concentrated in a fixed location, yet also needs to be spatially mobile in order to realize its value and avoid overaccumulation crises—the premise of the material fix is that this contradiction both shapes and is shaped by the people and material characteristics of the location in question. While this is particularly true in the case of resource extraction, in which local devaluation can be caused not only by crises of overaccumulation but also by degraded deposits, the material qualities of place matter even in less obvious contexts.

In Llallagua-Uncía both Patiño's private company and the state-owned COMIBOL struggled continuously with fluctuating global market prices, a large and politically active workforce, and a slowly degrading resource deposit. When either the ore grade or the market price of tin slumped, the companies responded with creative uses of three-dimensional space. Patiño's techno-spatial solution of block caving increased the rate of tin recovery, while COMIBOL's subsidiary organizations absorbed excess laborers and lowered the company's costs. The creation of mining cooperatives in 1987, following the closure of COMIBOL in 1985, might be imagined as the ultimate material fix since it evacuated the region of capital but pacified newly unemployed miners. This is what I mean by a *material fix*: the reorganization of nature, technology, and labor to stave off devaluation crises provoked by either market fluctuations or resource exhaustion.

Llallagua and Uncía are a palimpsest of material fixes that now mark the surface and subsurface of the towns. This three-dimensional palimpsest continues to divide local mining cooperatives that appear, at first glance, to share the same history. For instance, there is a noticeable tension between the cooperative Siglo XX, which was formed of ex-unionized miners and operates in the lowest, richest levels of the mountain, and the two cooperatives 20 de Octubre and Dolores, which were formed of ex-*locatarios* and operate in the highest, most exhausted levels of the mountain. Meanwhile, the two cooperatives that emerged from the *lameros*, Carmen and 23 de Marzo (which also absorbed most of the *palliris*), are the poorest in the region. The fixes that addressed the exhaustion of the mine and falling commodity demand have left socioeconomic rifts that are spatialized vertically within the mountain and horizontally across the surface: the material history remains as a material geography.

3

TANGLED VEINS:
OF TUBERS AND TIN

When dawn breaks over the peak of the Juan del Valle Mountain, miners are already congregating around the Cancañiri mine shaft. At an altitude of more than four thousand meters, Cancañiri is the highest of the three major mine shafts in Llallagua. As the miners unlock their shared storage garages and find their boots, helmets, and lamps, vendors open the doors of their kiosks to sell coca leaves, snacks, soda, and small plastic tubes of pure alcohol. Although it is barely 7 a.m., many miners are eating two-course lunches, since they will not eat again until they leave the mine in the late afternoon. The wind is biting as the smell of peanut soup wafts out from the canteen, drawing people toward it as much with the promise of warmth as flavor.

Most of these miners are wearing *k'epirinas*, compact backpacks made from potato sacks (see figure 3.1). *K'epirinas* are practical—miners use them to carry supplies underground and ore aboveground—but they are also material symbols of Bolivian cooperative mining. Such bags were less essential in the eras of privatized mining (1900–1952) and nationalized mining (1952–85), when miners shoveled ore into carts that were wheeled out with the help of electrical cables. Small-scale cooperative miners, in contrast, must haul *k'epirinas* full of crushed rocks through mazes of claustrophobia-inducing tunnels, up dozens of ladders, and along the mile-long main passageway before emerging outside.

The symbolic importance of the bag is also captured in written form: *K'epirina* is the title of a series of seven comic books published in Llallagua by the local radio station, Radio Pio XII, starting in 1987.[1] At a moment when mining cooperatives were just emerging, the comics attempted to clarify what these collectives were, how they were going to operate, and where they would fit within the national economy. As far as I know, the comics never circulated far beyond Llallagua, but within the town I found them everywhere: at the FERECOMINORPO (Regional Federation of Mining Cooperatives of Northern Potosí) offices, in the mining archives, in kiosks on the main plaza, and in the radio stations. Combining historical education, economic training, and entertainment, the comics humanized the experience of economic restructuring in the tin mines.

A critical read of *K'epirina*, however, also shows how indigeneity came to operate within mining cooperatives in the post-1987 period—and, by extension, how the *k'epirina* itself symbolized the "indigenization" of the mining cooperatives, which were increasingly populated by migrants from the ayllus of Norte Potosí. The comics follow the life of Pedro, a young Indigenous campesino from the ayllu Chullpa, which encompasses the land around Llallagua. Pedro is forced to leave his village when his crops are lost in a bad frost—a moment that corresponds to real-life events of the 1980s, when a series of droughts and frosts decimated regional produce. When he arrives in Llallagua wearing a *chulu*, a woven hat that marks indigeneity in this region, he encounters a friend whose own *chulu* is partially obscured by a miner's helmet (see figure 3.2). The friend invites Pedro to a rally celebrating the government's 1987 decision to inaugurate the mining cooperatives. Joining unemployed miners, students, and other recent migrants from the ayllu, Pedro enthusiastically embraces the chance to find his own fortune in the Juan del Valle Mountain. The comic nuances the typical narrative of laid-off miners becoming *cooperativistas* by showing how neoliberal policies, entangled with regional politics and weather patterns, also precipitated a movement from the ayllus into the mines.

These comics are unusual in that scholarly and popular discourses in Bolivia typically separate miners and members of ayllus into distinct and

FIGURE 3.1. *K'epirina*, the bag. Pictured at the Siglo XX mine shaft in Llallagua, September 24, 2016. Photo by author.

even antagonistic social groups. In economic terms, miners are framed as *obreros* (workers), while ayllus are predominantly agricultural and their members are often called *campesinos* (peasants or smallholding farmers). In racial terms, miners are usually interpolated as mestizos (mixed race, or more specifically non-Indigenous), while those from the ayllus are framed by a variety of labels that indicate indigeneity: *originarios* (original people), *autóctonos* (autochthonous people), *comunarios* (community members), and, pejoratively, *indios*. Finally, in spatial terms, miners are

FIGURE 3.2. *K'epirina*, the comic (excerpt). Pedro first hears about mining cooperatives. *K'epirina*, no. 1 (1987), 4.

imagined as dwelling in the urban zones of Llallagua-Uncía, rather than the surrounding rural areas. A similar point could be made about what is expected of an *originario*, a word that conjures not only the temporal prior but also landscapes unmarked by colonial activities. As discussed in earlier chapters, such places are not only rural but also surface-level; the subterranean is reserved for mestizo nationalism. Yet Pedro of the *K'epirina* comic migrates fluidly between his ayllu and the mine several times over the course of the comic's seven-part story.

Against the supposedly commonsense separation of the mines and the ayllus, most of the cooperative miners I met in Llallagua and Uncía were more like Pedro. They identified as *agro-mineros*, or agricultural miners. While mining is their primary source of income, they also have farmland that they tend at least a few times a year—to plant, to harvest, and to freeze-dry potatoes to preserve them through the winter. For at least half of the cooperative miners, this farmland is situated within one of several ayllus that surround the two towns, where their land rights also oblige them to participate in ayllu government. Cooperative miners even suggest that their individual connections to the land have enabled them to collectively overcome historical tensions between the two sectors (mining and agriculture). As a retired cooperative miner put it in a public lecture, "In the era of the unions, we talked about [needing to build] the *obrero*-campesino [worker-peasant] alliance, but now that work is being carried out *within* the cooperatives" (Llallagua, September 22, 2016). In other words, a tense and always external alliance between miners and campesinos from the early 1900s to 1980s had been internalized within the practices and identifications of individual cooperative miners. This statement was followed by raucous applause by the audience members, who imagined themselves bridging the divides that had troubled their predecessors.

Inspired by such moments, this chapter makes three interrelated arguments. First, after nearly a century of Indigenous exclusion from the tin-mining sector, the 1987 inauguration of mining cooperatives marked a moment when people from the ayllus of Norte Potosí began actively mining as well as farming. Second, this integration was important not only because it marked a moment of regional economic diversification but also because it constituted a local indigenization of the subterranean. In the Bolivian highlands, the socioeconomic categories of *obrero* and campesino were always already racialized, in that they were created from the substrate left behind by earlier colonial eras and racial projects, and the creation of mining cooperatives indexed a watershed movement of Indigenous people into a space that has been constitutive of mestizo nationalism: the subterranean tin mine. Third, the two objects the *k'epirina* would typically contain, potatoes and tin ore, are materially constitutive of the entanglement of mining and agriculture, and of indigeneity and *mestizaje*—that is, the ideology of racial optimization that undergirded many twentieth-century Latin American nation-building projects. While potatoes are an archetypal Indigenous Andean food crop, tin ore is a mestizo substance, both in that tin mining economically underpinned the institutionalization of mestizo nationalism after the 1952 National Revolution *and* in that tin mines have historically been treated as sites of mestizo formation. Tubers and tin thus constitute the tangled subterranean veins traced in this chapter.

Following the material histories of tubers, tin, and *k'epirinas* also helps explain contemporary dimensions of nation, race, and class in Bolivia. In particular, the way that indigeneity has been articulated within mining cooperatives is key to understanding expressions and rejections of resource nationalism within Bolivia's plurinational state. Although other informal sectors, such as El Alto's Aymara merchants or Cochabamba's *cocalero* (coca growers') unions, are similarly reworking notions of indigeneity, cooperative miners' subterranean labors are particularly transgressive. Operating in the liminal space between ayllu and nation, cooperative miners wind up shaping both local Indigenous politics and national economic policies, ultimately redefining what it means to be on the political left in the contemporary moment. All in the same sack, potatoes and tin are shaping politics in unpredictable and not always legibly progressive ways.

RACE AND INDIGENEITY IN BOLIVIA

While many scholars have noted that the solidification of class-based identities in the Bolivian tin mines coincided with the emergence of modern nationalism from the 1920s to the 1950s, few have had much to say on the topics of race or racialization in the tin mines. This is especially surprising given that mining has often been imagined as a quintessentially modernizing industry, an attribute that has been associated with the racial "improvement" of its workers.[2] Much like the military, Bolivia's tin mines were conceived as sites of alchemical transformation: separated from their lands, languages, and modes of production, *indios* could become mestizos in the disciplinary space of the mine (Kohl, Farthing, and Muruchi 2011).

This section and the next trace a brief history of race and indigeneity in Bolivia and Norte Potosí to argue that the categories of campesino and *obrero*, while apparently class based, both have their origins in colonial categories of race; for this reason, the incursion of the *agro* into the world of the *minero* also marks a shift in spatial delineations of race and indigeneity. These two terms—*race* and *indigeneity*—are not equivalent, but I follow other scholars in insisting that, in Bolivia at least, they are indivisible.[3] While settler colonialism (as exemplified by, if not limited to, contexts such as Canada, the United States, and Australia) is often understood as a structure of domination that seeks to expropriate land rather than exploit labor, this contrast is harder to discern in much of Latin America.[4] Spanish and Portuguese colonizers brought enslaved Africans to labor in Latin America, but they also enslaved Indigenous peoples and—after Indigenous slavery had been outlawed—continued to exploit them through mechanisms like the *mita* (forced labor in mines, framed as tribute to the state) and haciendas (plantation-style farms where Indigenous people performed forced labor as a tax on land rights).[5] This was particularly true in the Andes, where most (not all) colonial and postcolonial mining was undertaken by Indigenous peoples. Moreover, instead of a "one-drop rule," Spanish and Portuguese colonial bureaucracies enumerated a proliferation of racial mixtures known as *castas* (castes). White, Black, Indigenous, and sometimes Asian genealogical contributions were incorporated into the *casta* system. Although this system was far from a racial democracy, in that each new category was shored up as a new kind of "pure mixture" that reinforced rather than challenged racial hierarchies, the process of creating and administering this multiplicity nevertheless blurred the boundaries between race and indigeneity (de la Cadena 2000; Wade 2017).

This history is reflected in the contemporary political aspirations of Black and Indigenous Latin Americans. For instance, Afro-Latinx communities in places like Nicaragua, Honduras, and Brazil have fought for collective territorial control, a right that is typically associated with indigeneity in the United States (and in some other parts of Latin America) (see Hooker [2005], Mollett [2006], and French [2009] on Nicaragua, Honduras, and Brazil, respectively). Meanwhile, Indigenous peoples in places like Bolivia and Ecuador often frame their political struggles in the language of autonomy rather than sovereignty, a lexical distinction that indexes an interest in transforming, rather than completely exiting, the structures of the colonial nation-state.

In Bolivia indigeneity also operates very differently depending on whether one is considering the highlands, the lowlands, or the intermediary valleys. The western highlands and the intermediary valleys are the traditional homelands of the Aymara, Quechua, and Uru peoples. This region was part of the Incan Empire and was colonized by Spain comparatively early (in the sixteenth century). For the Spanish, the highlands were for mining, and the valleys were for agriculture, which meant that Aymara and Quechua ayllus in the highlands were forced to pay tribute and send *mitayos* (corvée laborers) to the silver mines, while Quechua communities in the valleys were forced to work as *colonos* (sharecroppers) on the haciendas of landed elites (Gotkowitz 2007). Meanwhile, Indigenous peoples of the lowlands—which include the Guaraní, Chiquitano, Guarayo, and Ayoreo, as well as numerous other groups—were brought under Spanish control via Catholic missions established over the course of the seventeenth and eighteenth centuries (Postero 2007). This distinct history remains preserved in the words used to discuss Bolivian indigeneity. While the word *indígena* is used by lowland communities, *originario* is more often used by members of highland ayllus, while Quechua speakers in the valleys often identify more readily with the socioeconomic category of campesino.

How to translate this distinction into English-language academic analyses is often a question in Bolivian ethnographies. In his ethnography of a highland Aymara community, anthropologist Andrew Canessa (2012, 7) uses the word *indian* (following Aymara intellectuals Fausto Reinaga's and Felipe Quispe's use of the word *indio*) because the word "indigenous . . . can obscure the long history of oppression of one people by another." Working across Ecuador, Peru, and Bolivia, Mary Weismantel similarly uses the word *indian* because avoiding it would mean indulging in an evasion of power that "allows whites to avoid confronting their own discomfort with questions of race" (2001, xxxiii). I appreciate this argument but hesitate to use a word that I heard less than a half dozen times in Llallagua-Uncía, always from the mouths of light-skinned business owners criticizing recent migrants from the ayllus. By the same token, the polite words *originario*, *autóctono*, and *comunario* seem devoid not only of racial content but also of colonial history, or the degree to which indigeneity was constituted within global processes of subjugation and subjectification. Given this predicament, I have opted to use the capitalized English word *Indigenous* when speaking about people in Norte Potosí who identify as *originarios*, a move that is intended to gesture toward discussions of colonialism beyond as well as within Bolivia.

Throughout the highlands and valleys, the colonial-racial term *indio* was both a fiscal category, which corresponded with a tributary regime, and a census category, which corresponded with a separate legal-juridical system, distinct from the one used for Spaniards and their descendants (Harris 1995; Wade 1997). But these systems of fiscal and legal separation were complicated by ongoing processes of cultural and sexual mixing, sometimes consensual but more often underpinned by violence. At the turn of the twentieth century, elites used the disparaging term *cholo/chola* to describe both people of mixed ancestry and people of Indigenous descent who inhabited racially liminal places, such as marketplaces or mines (Rivera Cusicanqui 1997; Weismantel 2001). Although the figure of the *chola*—recognizable by her many-layered *pollera* skirts and bowler hat—has since been "domesticated" as a national symbol, elites initially perceived *cholaje* as a threat to political stability and contrasted *cholos* unfavorably with "pure" *indios*.[6]

The values assigned to these two categories of *cholo* and *indio* began to shift, however, with the growth of discourses of *mestizaje* in the early twentieth century. As described in the introduction, proponents of *mestizaje* argued that miscegenation would be politically equalizing and would function as an antidote to colonial obsessions with blood purity. What *mestizaje* boiled down to, however, was a reassertion of a "mixed purity" that, while it might have valued Indigenous and Afro-Bolivian genetic contributions to a shared nationhood, did not value existing Indigenous or Afro-Bolivian communities, not to mention their respective political pursuits (de la Cadena 2005). Nevertheless, discourses of *mestizaje* were folded into the center-left political party that took power after the 1952 National Revolution, where it informed a postrevolutionary decree that all Bolivians were racially mestizo, separated from one another only by economic and "cultural" differences.

The new category of campesino described precisely the same populations that had once been *indios* and essentially rewrote a racial split in economic and spatial terms (Burman 2015). The term *campesino* gained currency with the 1953 Agrarian Reform, which encouraged the formation of peasant unions, and with the 1979 formation of CSUTCB (Unified Syndical Confederation of Peasant Unions of Bolivia). This was at the height of the so-called military-campesino pact, in which a series of military dictators pledged to support campesinos' unions in exchange for political support against the (increasingly radicalized) workers' unions (Dunkerley 1984). The CSUTCB was and remains strongest in the valleys, where most Indigenous peoples had been subjected to hacienda agriculture rather than *mitayo* labor in the mines.

As a regional and more explicitly Indigenous counterpoint, in 1997 Indigenous highlanders formed CONAMAQ (National Council of Ayllus and Markas of Qullasuyu). During the nearly twenty years that passed between the formation of CSUTCB and CONAMAQ, rural political organization was increasingly articulated in the language of Indigenous territorial autonomy rather than peasant land rights, a transition led in part by the militant La Paz–based *kataristas* (see the introduction for this history). But the distinct orientation was also a result of the divergent colonial history of the highlands, where ayllu governance was less completely obliterated by Spanish rule than it had been in the valleys.

Nowhere was this truer than in Norte Potosí, where campesino unions were layered on top of ayllus, both of which have more recently infiltrated the subterranean via mining cooperatives.

NORTE POTOSÍ: AYLLUS, UNIONS, AND COOPERATIVES

The mining towns of Llallagua and Uncía are located within or near three important ayllus. Llallagua is bordered on all sides by the ayllu Chullpa, while the two most influential ayllus on the Uncía side of the mountain are Kharacha and Aymaya.[7] I was first introduced to Chullpa by Mauricio, a leader of the mining cooperative Dolores whom I had met at the FERECOMINORPO offices. He was born and raised in Chullpa, and he invited me to accompany him to the peak of the Juan del Valle Mountain on his motorcycle so he could show me the ayllu lands below.

It was February, the rainy season, and the dirt path that wound its way up the side of the mountain was treacherous with stones emerging from the mud. But we persevered, dismounting and pushing the motorcycle over rough patches. As we climbed, we could see increasingly more of the land around us: miles of rolling hills, dotted with villages and potato fields. Gesturing in the general direction of his village, Mauricio told me about his childhood as he navigated up the mountain. As a boy, he had traveled seasonally by foot between his village high in the altiplano and the much lower valley regions, where he and his father traded potatoes and llama wool for grains, fruit, and vegetables. It was not an easy business. Mauricio described waking up in the open air having been covered with snow in the night, and the trials of trying to find water for the llamas in the dry season on the altiplano. Although it was only a few hours' walk from the mining town of Llallagua, few people from Mauricio's village became miners. "My father was of the generation that was scared of the mine," Mauricio commented. "And there was a lot of discrimination back then, you know. Not just anyone could work in the mines. They didn't hire from the area" (Llallagua, February 28, 2016).

Historians of mining in Llallagua-Uncía have made the same observation. Luis Oporto Ordóñez's (2007) study, which focuses on the early era of tin mining in Llallagua-Uncía (1900–1935), argues that workers in Uncía came largely from the exhausted silver mines of Colquechaca, within Norte Potosí, while workers in Llallagua came largely from the valleys of Cochabamba and Oruro. Another wave of workers came from beyond Bolivian borders, with a particularly high number of Chileans who had been working for the Chilean Compañía Estañífera de Llallagua (Tin Company of Llallagua, which was later purchased by Simón I. Patiño, the prospector introduced in chapter 2). Oporto Ordóñez did not find a single case of a worker born in Uncía prior to 1913 (24). My own review of 300 randomly sampled employee files in the COMIBOL Archives, all associated with workers hired between 1910 and 1929, confirmed his results. Within this sample, 110 workers (37 percent) came from the department of Cochabamba, 55 (18 percent) were from the department of Oruro, 46 (15 percent) were from the department of La Paz, and 23 (8 percent) were from other parts of the department of Potosí. Although 41 workers (14 percent) were from Norte Potosí, nearly half of these were from Colquechaca, and only 22 (7 percent) were from villages and ayllus.[8]

It is impossible to know exactly why Patiño limited his recruitment activities in the immediate surroundings. When I asked Llallagueños, many hypothesized that Patiño, himself a Cochala (from Cochabamba), hoped to surround himself with familiar foods, language, and cultural practices. My own best guess is that he was looking for people whose dispossession was more complete. Not only were the descendants of Cochabamba colonos accustomed to brutal forms of exploitation, but they likely also wanted to leave the haciendas; hence, they sought employment in Patiño's tin mines. By contrast, the colonial "pact of reciprocity," which had granted the ayllus of Norte Potosí relative autonomy in exchange for tributary labor (Platt 1982), had also allowed ayllu members to maintain their distance from the colonial economy's wage relations. Patiño, with his vision of modernization at the turn of the twentieth century, created a classic enclave economy in the heart of Indigenous territorial control.

Even the 1952 National Revolution, which resulted in the expropriation of Patiño's mines and the creation of COMIBOL, had very little impact on this sociospatial division. While the system of campesino unions was embraced in the valleys, the ayllus remained the dominant form of rural organization in Norte Potosí. Although residents of Llallagua-Uncía might have described their rural neighbors as campesinos, they did not organize in ways prescribed for campesinos—theirs was a rotational rather than a union-centered organization. This reality began to shift in 1983, in the context of a devastating drought, when a local NGO (nongovernmental organization) insisted that the rural residents of Norte Potosí organize themselves into unions to receive food and cash relief. Counterintuitively, the imposition of unions wound up strengthening the ayllus, since it was through the former that the *katarista* movement took hold in the region. The La Paz–based group of *katarista* anthropologists who arrived in the mid-1980s to strengthen Norte Potosino ayllus wound up partnering with the local NGO that had helped establish the unions. The anthropologists' conversations with the new campesino union leaders helped spark a regional *katarismo*, this time based on the importance of regional hero Tomás Katari rather than national hero Túpac Katari, and paved the way for the creation of CONAMAQ in 1997 (Choque and Mamani 2001; Rivera Cusicanqui 1992). At the same time that the miners' unions of Llallagua-Uncía were being dismantled by the onslaught of neoliberal economics, the surrounding ayllus were increasingly at the forefront of Bolivian Indigenous activism.

Most scholars would agree that the mines and ayllus of Norte Potosí were artificially divided, in that there was always some interaction between the two zones despite their separate political, economic, and even linguistic systems. Less commonly observed, however, is that the neoliberalization and virtual closure of the state mining corporation COMIBOL in 1985 significantly weakened this always somewhat permeable barrier. When mining cooperatives were first formed in Llallagua in 1987, ex-salaried miners enlisted the help of migrants from the ayllus to help them extract as much as possible as fast as possible, as they did not expect the cooperatives to be a durable economic formation.[9] Like Pedro from the comic *K'epirina*, this is how Mauricio's father first began working in the cooperative Dolores. When his father died from a severe stomach infection, Mauricio, age sixteen, took up his mantle in the mine. But in his *cuadrilla* (work crew) of thirty people, Mauricio distinguished himself with his mathematical skills, quickly taking over the *cuadrilla*'s financial operations rather than spending his time underground. He began occupying leadership positions within the cooperative, and at age thirty-six he became the cooperative president, a feat that he accomplished while returning on a weekly basis to tend to his land and participate in ayllu government.

Although the descendants of salaried miners arguably still hold the most power within cooperatives, stories like Mauricio's are increasingly common. Many Aymara speakers (most of whom have also learned Quechua and Spanish in town and at school) have worked their way up through the ranks of the cooperatives, particularly the cooperatives that used to be subsidiary organizations (see chapter 2). In other cases, members of ayllus have forced their way into cooperatives en masse. Most dramatically, in 2005 a group of people from the ayllu Chullpa who had been working as day laborers for members of the mining cooperative Siglo XX organized a subterranean occupation. Entering at night, they barricaded themselves inside the mountain, taking all the tools and declaring themselves owners of the mountain, as it was situated within their traditional territory. After several days of armed conflict, the situation was resolved when Siglo XX accepted a hundred people from the ayllu into its ranks, granting the newcomers access to the lowest (and richest) levels of the mountain. They, like all the cooperative miners who return periodically to tend to agricultural plots, use the word *agro-minero* to describe their livelihood practices, but the racial transgression implied by this term is perceptible in light of the preceding history.

At a more material level, this transgression can also be read through the *k'epirina* and the objects it typically contains. *K'epirinas* moved into the underground along with miners from the ayllu, when a sack that had once been used to carry potatoes was converted into a backpack for tin ore. Potatoes have always been a primary source of food for miners in Llallagua and Uncía, but miners themselves were rarely involved in potato cultivation prior to the formation of mining cooperatives.[10] Now that there is more individual circulation between the mines and the fields, the symbolic content of the potato has gone deep underground with the miners themselves.

POTATOES IN A SACK

As we rode his motorcycle back down the Juan del Valle Mountain, Mauricio told me two things that caused me to reflect on the racialization of the potato within this region.[11] First, he explained that when his father began working in mining cooperatives, his father and the other recent migrants from the ayllus were called *mak'unkus*, the Quechua word for the small green fruit produced by potato plants. This word marked the new arrivals as outsiders, even though the spatial and socioeconomic distance between miners and campesinos had been traversed. Second,

Mauricio pointed out the mountain from which the town of Llallagua takes its name. The Quechua word *llallagua* was translated for me as "conjoined twin," but it usually refers to potatoes that look conjoined (rather than conjoined people or animals). A *llallagua* potato is good luck, and this particular mountain is shaped like a *llallagua* potato, with two peaks rather than one.[12] Locals refuse to mine this mountain, although Mauricio assured me it is full of tin. As he explained, the mountain would give up its ore only to those willing to leave a human sacrifice, and none were quite so eager for its riches.

I began considering the potato—soil bound rather than deeply subterranean—as a regional analytic counterpart to tin, the metal whose wayward itineraries shaped the mining towns of Llallagua and Uncía. Potatoes, after all, are perhaps *the* quintessentially Andean crop.[13] The potato was first domesticated some eight thousand years ago on the shores of Lake Titicaca, and most potato varieties have been bred to withstand high altitude, intermittent drought, and freezing temperatures. Because of this hardiness—not to mention tastiness—they remain an integral part of almost all highland Bolivian meals.[14]

Yet, thanks to colonial trajectories, the potato is also inextricable from a globally constituted web of meaning. On first encountering the potato in the mid-1500s, Spanish conquistadors returned to Spain with boatloads of the tubers, delighted that they had both a long shelf life and enough vitamin C to stave off scurvy.[15] The middle- and upper-class Europeans who encountered the potato at home, however, were less enthused. The potato, European chefs agreed, might work for semibarbarous sailors but was hardly fit for upstanding citizens. Reports from the time emphasize the potato's mealy texture and indigestibility, even accusing it of infecting people with diseases ranging from leprosy to syphilis. Their hesitation might have been warranted if cooks unfamiliar with potatoes had tried to serve the plant's leaves, which, like all members of the Solanaceae (nightshade) family, are mildly toxic (Reader 2008).

Despite this formal rejection, however, European peasants quietly went about planting their fields with the hardy tuber, hedging against poor grain harvests. Like the sailors before them, they found that potatoes offered a long-lasting and relatively nutritious alternative to wheat and oats. Thick discursive vines grew between potatoes and peasants, neither of which were popular within elite European society. Or, at least, European elites did not care to eat the potato themselves; they were happy to feed the cheap but calorie-dense tuber to peasants and the emerging working class. As historian Redcliffe Salaman argued, "The potato can, and generally does, play a twofold part: that of a nutritious food, and that of a weapon ready forged for the exploitation of a weaker group in a mixed society" (1949, 600). Most famously, the tuber became such a staple in the Irish diet that when the fungus *Phytophthora infestans* began destroying Irish potato crops in 1845, over a million Irish died of starvation and related causes.

Ireland had barely recovered from the potato famine when Karl Marx penned *The Eighteenth Brumaire of Louis-Napoleon Bonaparte* ([1852] 1963), an influential study of how political consciousness can fail to form, particularly among peasants. Lamenting the results of the 1848 French election, which returned the country to autocratic rule, Marx blamed the French peasants, who had failed to develop collective class consciousness because their access to land—however measly their plots—protected them from market vacillations and isolated them from one another. Marx's interpretation of peasant politics was nothing short of scathing:

> A small holding, a peasant and his family; alongside them another small holding, another peasant and another family. A few score of these make up a village, and a few score of villages make up a Department. *In this way, the great mass of the French nation is formed by the simple addition of homologous magnitudes, much as potatoes in a sack form a sack of potatoes.* In so far as millions of families live under economic conditions of existence that separate their mode of life, their interests and their culture from those of the other classes, and put them in hostile opposition to the latter, they form a class. In so far as there is merely a local interconnection among these small-holding peasants, and the identity of their interests begets no community, no national bond and no political organization among them, they do not form a class. They are consequently incapable of enforcing their class interest in their own name, whether through a parliament or through a convention. *They cannot represent themselves, they must be represented.* (Marx [1852] 1963, 124, emphasis added)

Much ink has been spilled exploring, refining, and refuting this passage (Hobsbawm 1973; Shanin 1982). It has been used as evidence of Marx's thoughts on the political consciousness of peasants, or what Henry Bernstein (1996) has called the "political dimension" of the agrarian question.[16] As potatoes in a sack, peasants are unable to grasp or act on their collective political interests and can therefore derail the workers' revolutionary efforts. Although Marx wrote many nuanced reflections on the relationship between capitalism and agriculture, his strong statement about the lack of political consciousness among peasants has nevertheless infuriated generations of politically progressive smallholders.[17]

Rarely does anyone note, however, that in 1852 Marx seemingly did not find it strange to speak about potatoes as if they were the archetypal food of the French countryside, when they were in fact evidence of a relatively recent colonial connection with the Andes. Potatoes were actually banned in France until the 1770s, when an army physician proved definitively that the tuber did not cause leprosy (Reader 2008, 120).[18] Whether he knew it or not, Marx's analysis of the French peasantry was steeped in colonial history. The relational history of knowledge production informed not only the vegetable metaphors that Marx used to discuss the French countryside but also global associations between peasantry and potatoes: both poor and tasteless, no matter where in the world they were connected.

Of course, Marx's conceptualization of peasant politics can also be considered within the broader history of racialism in Europe, which—as Cedric Robinson ([1983] 2020) argues in the early chapters of *Black Marxism*—preceded the emergence of capitalism and influenced the formation of the European working class. The racializing discourses about the Irish by Anglo-Saxons are particularly familiar to contemporary race scholars, in part because they lasted so long; as Anne McClintock (1995) notes, representations of the Irish in the nineteenth century often depicted them with the same "barbarous" features used in discussions of African and Asian peoples, creating a common colonial race hierarchy. By situating Marx's analysis of French politics within Robinson's history of European racialism, one reaches a different conclusion about apparent class conflict. When Marx was writing, the French merchant class (mostly Celtic Gauls) was pushing back against the landholding Franks in the newest iteration of a conflict that had lasted centuries (Foucault [1976] 2003).[19] Yet the *way* that Marx critiqued French peasants was also tangled in much more recent colonial threads. The French potato indexes a growing intimacy between racialized class politics in Europe and racialized colonialism in the Andes.

Although potatoes are a dish of choice for contemporary Bolivian highlanders across the socioeconomic spectrum, growing potatoes is—as in nineteenth-century France—associated with rural, small-scale agricultural communities. And, as in France, these communities are often regarded with suspicion by many Latin American Marxists, particularly those immersed in the Trotskyism for which Bolivian miners' unions were so well known. But I am less interested in demonstrating campesinos' political consciousness than I am in considering the extraeconomic content of the Marxian category of the peasant.[20] By calling them potatoes in a sack, Marx was racializing (with a colonial referent) as much as he was distinguishing by livelihood. The global movement of potatoes as a plant has been matched by a global circulation of discourses that relegate the plant to rural/surface landscapes, where they become entangled with local histories of colonial race hierarchies and rural poverty.

The very best potatoes I have eaten were served after a full day of *chuño* production in the outskirts of Uncía. Prepared in the coldest month of the year, *chuño* is made by freezing potatoes overnight, removing their skins, and leaving their flesh to bake in the sun. Removing the skins is a pedal process; no machine has proved as efficient as repeatedly rolling the defrosted potatoes beneath one's bare feet. It is also an occasion. Every year in early June, the towns of Llallagua and Uncía clear out as residents return to family lands to make *chuño*. In 2017 I was invited to assist in a *chuño* stomp by the family of Tata Max, then the leader of CONAMAQ, who was from the ayllu Aymaya but had been living in Uncía for nearly a decade. We celebrated the completion of the day's work with *watía*, a feast centered around potatoes baked in an underground oven—an excellent antidote to sore leg muscles and frozen toes. *Watía* potatoes smell like fire and earth and need neither salt nor spice to taste divine.

I was therefore taken aback when Tata Max apologized for not having any meat to serve. "This is why we need help identifying metals in our ayllu," he informed me, and went on to describe his dream of using his ayllu's natural mineral wealth to enrich its residents. Tata Max's vision pushed back against the legal and discursive ropes that bound the ayllu to land-based activities like farming, but this desire to exceed the category was also driven by dissatisfaction with what the land could produce. Colonial double binds: demanding the subsoil both subverts colonial cartographies that limit indigeneity to the topsoil *and* tacitly accepts a hierarchy of material objects in which tubers are figured as less desirable than metals. This hierarchy, established through the colonial circulations of potatoes from the Andes to Europe and back again, is not only racialized but also racializ*ing* for the people who cultivate those tubers today.

Llallagua, like most mining towns across the Bolivian highlands, is home to several statues of miners. As in the other towns, Llallagua's statues materialize ideals about what miners are supposed to look like. Male presenting, tall, brawny, and shirtless, usually holding the tools of Bolivia's imagined modernity: a drill in one hand and a gun in the other. Mouth open in a revolutionary cry, the statue in Llallagua's Siglo XX Plaza has his gun raised in battle (see figure 3.3). He is helmeted but otherwise unprotected from whatever he is confronting. His flat abdomen suggests a poverty of potatoes.

These statues appear racially unmarked, but this is another way of saying that they are identifiably mestizo. They are not wearing any of the clothes that signal Andean indigeneity, such as a *chulu*, poncho, or *pita* (the rope that Quechua and Aymara men sometimes wear across their chests). Instead, the statues are clad in jeans, boots, and helmets. This marked unmarkedness contrasts to another statue that often appears as the standing man's sedentary complement: a *palliri*, or a woman who searches for valuable ore left in discarded rocks on the surface of the mine. The *palliri* statue uses a hammer to manually crack rocks open and is dressed in a bowler hat, a full pleated skirt (*pollera*), and a shawl pinned around her shoulders (*manta*)—in short, the symbols of a *chola* (see figure 3.4). Her femininity and *cholaje*, as well as her figure's apparent passivity, enhance the action of her companion—the masculine, mestizo tin miner who is the protagonist of twentieth-century Bolivian nationalism.

These statues embody a particular set of meanings that condensed around tin miners and the substance of tin in the early twentieth century: growing trade unionism, an increasingly widespread sense of national unity based on *mestizaje*, and a desire to wrest economic control over natural resources from local elites and imperial powers. When the tin-mining sector was booming, it was framed as the key to economic independence and the missing ingredient of national sovereignty. In sharp contrast to peasant farmers of potatoes and quinoa, who were apparently concerned only with their own household economies, tin miners were seen as literally digging the nation out of the dependency cycles that had plagued Bolivia since winning political independence in 1825. Tin was a national matter, a working-class affair, and the harbinger of mestizo masculinity. Tin mines, meanwhile, were mestizo-producing

FIGURE 3.3. Statue of a miner in Llallagua's Siglo XX Plaza, April 1, 2017. Photo by author.

machines, capable of transforming Indigenous campesinos into Bolivian workers fighting for the nation.

The attribution of such alchemical qualities to tin mines stemmed in part from the racial ambivalence of the subterranean within the segregated spaces of colonial and republican Latin America.[21] If the horizontal plane of the nation had been divided into criollo and *indio* spaces, the messy subterranean belied this neat separation. In addition to being far from cities and operated almost entirely by *indios*, the work of mining took place on the literal underside of national territory. But the matter of tin also lent itself particularly well to the project of *mestizaje*. Un-

FIGURE 3.4. Statue of a miner and *palliri* in the city of Potosí, August 7, 2013. Photo by author.

like silver, the iconic metal of Spain's colonial economy, tin is far from a symbol of elite decadence. No one makes jewelry, coins, or religious iconography out of tin. In fact, the absence of silver and gold in Norte Potosí offered some protection against colonial incursions in the 1600s: the Juan del Valle Mountain is named after a wandering Spanish conquistador who moved on after failing to find silver in the mountain's folds. By the time Patiño was prospecting in Llallagua nearly three hundred years later, however, tin had become a key metal of industrial modernity. In the Global North, particularly during the tin shortages of World War II, the metal was associated with factories, soldiers, and US housewives serving dinner out of tin cans. In Bolivia tin's industrial association linked it to senses of nationalism, dreams of economic sovereignty, and *mestizaje*.

As with the statues in Llallagua, these links between tin mining and mestizo nationalism often show up in public art. In June 2016, while I was helping Don Alfonso (see chapter 2) and the rest of the FERECOMINORPO directorate paint the cement risers of a new sports stadium in Llallagua, I happened to meet the artist who had designed and built Llallagua's miner statue. Iván, Llallagua's resident artist, had been hired to paint the crests of each of the sixteen cooperatives that belonged to FERECOMINORPO along the inside walls of the stadium. A thin man in his sixties, Iván chain-smoked as he painted, the curls of smoke mixing with paint fumes as they circled up to the distant ceiling. When I paused to admire his work, he immediately engaged me in a long conversation about realism and politics in public art. He pulled out his phone to swipe through pictures of his other paintings, many of which were studies of men's muscles in action: close-ups of arms, chests, and backs. Inspired by Soviet realism, Iván's pieces more closely resembled action figures than any of the miners I had met.

Iván caught my attention, however, when he informed me that he was also the illustrator for *K'epirina*, the series of comics about Llallagua's mining cooperatives. He explained that he had spent six months working underground to really get a feel for the bodily experience of the labor before he began to draw. When I returned to my stack of *K'epirina* comics later that night, I saw the images with new eyes now that I had met their illustrator. It is truly a tragic tale. Pedro, the campesino-turned-*minero* who is the protagonist of the stories, spends many scenes deep underground reflecting on the sacrificial qualities of his work. He ultimately dies in an underground accident, and his widow cries over his grave while wondering how she will feed their children. As in most Bolivian literature, music, and art about tin mines, the theme of sacrifice is pervasive. Iván's drawings clearly aimed to elevate the workers' movement by depicting their adversity in such visceral ways, but in doing so he flattened miners' identities into a masculine warrior-worker, no longer marked by the clothing that, at the beginning of the comic, had signaled Indigenous identity. Dying in a tin mine was perhaps the ultimate mestizo action. The violence here is doubled: only through literal death was the transformation of campesino to mestizo complete.

These comics, although they reveal the connection to the ayllus that animates contemporary mining cooperatives, are also structured by a much older narrative in which tin mines act as crucibles of Bolivian nationalism, as articulated with both *mestizaje* and the subterranean. As discussed in earlier chapters, this connection among nation, *mestizaje*, and the subterranean was forged during the Chaco War (1932–35), in which Bolivia and Paraguay fought over the Gran Chaco desert, a territory rumored to be rich in oil. Bolivia lost the war spectacularly, but a sense of nationalism was nurtured in the trenches—a nationalism that was, as historian Kevin Young (2017) has argued, always foundationally a resource nationalism. This means that it was forged around the question of who should be benefiting from the country's natural resources. But resource nationalism, I contend, would not have been so strong if it were not also articulated with mestizo nationalism. The sense of shared ownership relied on a sense of national kinship. By this logic, if all Bolivians belonged to the nation through their shared mestizo kinship, then all Bolivians deserved to profit from resources contained within the national territory.

Unionized tin miners gave voice to a vision of nationhood in which they—as working-class men and mestizos—could benefit from the subterranean as national inheritance. This project demanded that they continuously distance themselves from all that was marked as Indigenous, and tin miners worked to consolidate their identities as above and against Indigenous campesinos. This boundary required constant policing. For instance, when anthropologists Olivia Harris and Xavier Albó (1976) argued that most tin miners in Norte Potosí were campesinos seeking wages to supplement rather than replace their farming livelihoods, celebrated union leader Filemón Escobar (1986) wrote a testimonial directly refuting their work, narrating instead a process of assimilation, or migration to the tin mines as a one-way trip to proletarianization. In my reading, the vehemence of his writing is a testament to the regional importance and perpetual fragility of the worker/peasant boundary.

This commitment to mestizo resource nationalism remains in the mining cooperatives' older guard, or miners who used to work for the state corporation before it was shuttered. For instance, when I was looking for information about the 2005 occupation of the Juan del Valle Mountain by the ayllu Chullpa, I was directed to speak with Ruben, who had been president of the mining cooperative Siglo XX at the time of the event (and who had since retired to drive a taxi). This interview not only revealed some of the ongoing rancor that the older generation of cooperative miners hold toward newcomers from the ayllus but also expressed tensions around the question of *how* claims to the subsoil are made, via what kind of inheritance:

> We the *cooperativistas* were assaulted by the ayllus, they treated us like thieves. . . . They said that they were the owners of everything, the mine, the land, and everything that is common [*llano*]. . . . According to their history, the Chullpas, those from the ayllu Chullpa, were the first ones in this town. So this has made them believe that as descendants they are owners of Llallagua. . . . I think that one way or another we wanted to tell them that *we are also born here*, in this town. We don't know much about the trajectory of our parents, but look, for example, I was born here in Llallagua, I am also equal, owner as much as anyone, no? I began my life here, my parents were miners, I was a miner. *They were never miners, the ayllus have never been miners. They were dedicated to the land.* But in some moment maybe they saw the money that can be made in the mine, and maybe they no longer wanted to work in agriculture and came here to put themselves in the mine as well. We never asked anything from them, and they wanted to force us out. I think that we were trying to make them understand that *no one is the owner here*, that they can also become *socios* [members of the cooperative] and they can also enter the mine and work, but they can't displace anyone. (Llallagua, July 9, 2016)

According to Ruben, while the migrants from the ayllus claimed the subsoil as an ancestral right, the descendants of salaried miners claimed it by virtue of their own individual births and their multigenerational labor underground. Ruben's last sentence also references a key aspect of cooperative miners' sense of collectivity. Rather than trying to stabilize the boundaries of the cooperative and redistributing wealth within it, cooperative miners insist on "free association," by which they mean that there are no restrictions on who or how many people can join. All Bolivians as Bolivians have a stake in the nation's patrimony, they reason, and should therefore be permitted to mine. This conviction is a distorted inheritance from the unions, whose commitment to national unity included a vision of shared access to the benefits of natural resource extraction, though unionized workers hoped these benefits would be diffused through a worker-led state rather than amassing to individual miners. Indeed, this is why tin miners so famously fought for the nationalization of the tin mines. As explored in chapter 1, this vision of subterranean nationalism is both unifying and deeply colonial, as it has been discursively subtended by an imaginary of *sub terra nullius* and historically produced through laws that barred Indigenous people from extracting subterranean resources. The cooperative redeployment of resource nationalism as an individual right to the subterranean is hardly what the unionists fought for, but the two discourses are kin in that they are grounded in nationalist belonging and descent.

In part because of the boundary work done by miners like Ruben and Filemón Escobar, the local genealogies of potatoes and tin are usually held apart. Yet the assumption that Indigenous peoples are not miners, while never historically accurate, is increasingly absurd in the contemporary tin mines. Moreover, it is inadequate for understanding how mining cooperatives conduct themselves as workers and political subjects in the contemporary era.

VERTICAL FARMING AND INDIGENOUS AUTONOMY

Although he is a cooperative miner from Uncía, I first met Carlos in Cochabamba, during the Twenty-Fourth National Congress of FENCOMIN (National Federation of Mining Cooperatives of Bolivia) in 2016 (an event detailed in chapter 6). From the moment I met him, Carlos was adamant that I accompany him underground because, he insisted, his working reality was different from that of other cooperative miners. Although they are members of the cooperative Siglo XX, Carlos and the rest of his *cuadrilla* labor in a strange corner of the mine, tucked away on Uncía's side of the mountain. Instead of entering through the major mine shafts (Cancañiri and Siglo XX), these miners had excavated an entrance of their own.

In June 2017 I rode with Carlos on his finicky motorbike up to this entrance, tipping over twice in minor accidents that he blamed on my being larger than the average Bolivian woman. These transportation troubles made us late, and by the time we arrived, the rest of his crew was already going in for the day's work. The two of us sat for half an hour in his *pauwiche* (storage and preparatory room) anyway, chewing coca and sipping from a bottle of Tres Plumas, a sweet coffee liqueur. Because they worked in short tunnels rather than long mine shafts, Carlos's crew used a small outdoor shack rather than an underground room as their *pauwiche*. Windowless and unlit except for a single hanging lightbulb, it might as well have been underground, except for the wind that whipped through the cracks in the sheet metal walls. As we chewed, Carlos outlined his life trajectory, explaining that he had migrated to Uncía from his hometown in the ayllu Kharacha to attend high school and had started working underground with his uncles shortly afterward. His *cuadrilla* was mostly made up of his brothers, his cousins, and a few close friends whom he described as part of his extended family.

My visit to Carlos's *paraje* (subterranean work area) was one of many that underscored the extent to which cooperative mining practices have been influenced by labor practices in the ayllus. Each cooperative is divided into hundreds of small work crews called *cuadrillas*, each of which has its own *paraje*. Much as *chuño* stomping had been a familial affair for Tata Max, *cuadrillas* are made up of fathers and sons, uncles and nephews, and even the occasional husband-and-wife team. The mining cooperatives have an official policy that disallows membership prior to age eighteen, but a collective blind eye is turned to teenagers assisting their parents, especially during school vacations. Indeed, given the current exhaustion of the mine, newcomers have a hard time finding a decent *paraje* without a family connection. Although becoming a member of the cooperative technically gives one the right to seek one's fortune anywhere within the concession, veins that have already been claimed are off-limits. Families stake out different sections of the underground and guard their findings with sheet metal, padlocks, and threats scrawled in spray paint.

Of course, mestizo miners likely also mobilized kinship networks to access jobs in the eras of privatized and state-run tin mining; the difference is not so much that kinship matters but that different forms of kinship matter much more than they used to. Belonging to an ayllu can now facilitate access to a subterranean *paraje*, whereas the opposite would have been true in previous eras. Moreover, the jobs acquired through mestizo relations were not as territorially segmented as those acquired through ayllu relations. While a father might facilitate his son's hiring in the era of mestizo mining, the two would not necessarily end up working in the same section of the mine. By contrast, kinship connections between the cooperatives and the ayllus tend to re-create familial "plots" underground, with each *paraje* populated by a set of people who also spend time planting and harvesting crops together on the surface.

Beyond this subdivision of vertical space, the divisions of labor within each *cuadrilla* also tend to mimic those of farming families. Tasks are distributed relatively equally and rotated to ensure that no one is always responsible for the most onerous ones. When Carlos and I finally strapped on our helmets and lamps and went underground, we walked straight to his *cuadrilla*'s single electrified room, which was the epicenter of the family's labors. It was more an enlarged section of a tunnel than a room, and the rocky walls were illuminated by a bulb that hung low from the ceiling. Salvador, Carlos's younger brother, glistened with sweat as he operated the air compressor, a huge cylindrical machine that sends air via pipes to pneumatic drills in unelectrified parts of the mine. Yelling over the rattle of the machine, Salvador explained that his cousin and childhood friend were down a ladder below, operating the drill deep in their *tope* ("limit," referring to the furthest point of extraction, where the vein is visible). Two people are required for this job, one to hold the drill and force it into the rock, and the other to direct the drill bit into the right place in the vein. Salvador went on to note that operating the air compressor was the easiest job and that it was his turn to work it. This was the cooperatives' rotational labor system in action. Unlike the hierarchical labor practices described in June Nash's ([1979] 1993) ethnography of unionized Bolivian tin miners in the 1970s, in which each miner had a fixed job that he performed every day, the *cuadrillas* within today's mining cooperatives rely on distributive labor principles imported from their experiences cultivating potatoes and other crops on their ayllu lands.

"Except it [the rotational labor system] doesn't work when people don't show up," Salvador ribbed Carlos, sparking guilty smiles. Salvador was not referring only to Carlos's tardiness that morning, though that likely didn't help. It had been a couple of weeks since Carlos had come up to the mine, as he had recently decided, after twenty years as a miner, to go back to school for a law degree. This meant that Salvador was next in line to inherit the *paraje*, so he had been taking on more leadership responsibilities. This is perhaps the aspect of the *cuadrilla* system that most closely resembles a family farm: *parajes* and their associated veins of ore are hereditary. While the sons of unionized miners were sometimes granted jobs when their fathers passed away, this inheritance did not come with any subterranean property rights. In the wake of the death of a *cooperativista*, by contrast, the vein he was exploiting typically goes to his brother or eldest son, although this gendered arrangement is increasingly flexible—for example, some widows now work in their husbands' steads. In decades past, heirs also had their membership fees waived, but that policy was updated with the commodity boom of the early 2000s. Now they must pay to join the cooperative, but once they are members, they have guaranteed membership in the deceased miner's *cuadrilla*. Therefore, although cooperative miners are often described as profoundly individualistic thieves, their productive capacities are enabled by kinship networks that turn *cuadrillas* into vertically organized family farms. This arrangement reflects the occupation of subterranean spaces by people who want to benefit from mining without identifying with the mestizo-national project.

The movement of agriculturalists into the subterranean has not been without friction, however, particularly given the material conditions of the mine. When ayllu members were incorporated en masse into mining cooperatives after the neoliberalization of the mining sector, they encountered a subterranean world in which social hierarchies were organized in relation to the geologically differentiated Juan del Valle Mountain, as described in chapter 2. This vertical differentiation implies different financial and health outcomes for *cuadrillas* employing the same labor techniques and same technologies. Such geological differences are much greater than the differences between agrarian families with different plots of land, particularly since lands can be made more productive with the additions of fertilizer, terracing, irrigation channels, and so on. Although miners can also improve their productivity levels with technological investments, these will never cause their *paraje* to yield more ore than it contains; instead, technology can increase either the rate of extraction or the percentage yield. Although the cooperatives have been shaped by the transposition of Indigenous agricultural practices into the subterranean, the material reality of the subsoil also means that these practices have different socioenvironmental outcomes underground. These activities don't "look" Indigenous in the ways that indigeneity is advertised within and beyond Bolivia—as socially complementary and in harmony with nature—but this does not signal capitalist corruption so much as the inadequacy of a category that is limited to land-based, agricultural activities.

More concretely, mining can also support Indigenous sovereignty in a way that agriculture is not always able to do. Mauricio, the miner from the ayllu Chullpa, described to me how he imagined this working in Norte Potosí. Standing on top of the Juan del Valle Mountain peering down at his ayllu, he explained that his long-term vision was to encourage members of his ayllu to think of mining as a community investment and to put part of their earnings toward the eventual establishment of a brewery that could rival Huari, the favored regional beer. The bulk of the ayllu, Mauricio noted, was situated above the mine and therefore did not suffer from downstream contamination; its waters were ideal for brewing. His vision echoed that of Tata Max, who was hopeful that mining might help fund his activities in CONAMAQ.

Although the desire to participate in extractive activities might strike progressive scholars as unsavory, a radical spatial politics is embedded in this claim. The surface—the land—is not enough for Indigenous sovereignty, particularly when the subterranean space can be rented by a foreign nation-state. Mauricio and Tata Max are both leaders in their communities, and by encouraging their friends and family to join or form mining cooperatives, they are strengthening the claim on the subsoil. In words and actions, they are disaggregating indigeneity and agriculture, giving shape to a racial counterformation that refuses surface-level circumscription.

After speaking with Salvador, Carlos insisted on taking me to his *tope*. I was in no hurry to go. Situated along a fault line, where cave-ins are frequent, Carlos's *paraje* was one of the few subterranean places that really scared me. I followed him reluctantly, scrambling over piles of barely stable rocks into a passageway so narrow that we eventually had to slither on our bellies. Carlos advised where not to put my weight, but it seemed like there was no place left where I *could* support myself. Rocks seemed to keep giving way under my hands and feet, making my heart pound as my boots filled with gravel.

Finally, we lay on our backs on a pile of stones, the ceiling barely a foot from our faces, and panted in the oxygen-poor shaft. Chuckling at my obviously panicked reaction to his *tope*, Carlos took the opportunity to talk more about his decision to leave mining and go to law school. He explained that his plan, on graduating, was to get into politics. He saw three possible avenues into national political arenas: through FENCOMIN (the national organization of mining cooperatives), through CSUTCB (the national organization of peasant unions), and through CONAMAQ (the national organization of ayllus). Mining cooperatives, ayllus, and campesino unions all govern different aspects of life for *agro-mineros* like Carlos, with the history of the miners' unions looming large behind all three. Regardless of the path he chose, Carlos's general plan was to rise through the ranks until he was able to join the COB (Central Obrera Boliviana; Bolivian Workers' Central), where he could represent *agro-mineros* alongside workers from all different sectors. Historically created and dominated by the FSTMB (Syndical Federation of Mine Workers of Bolivia), the COB in theory represents all workers in Bolivia, but its policies dictate that the leader must always come from within the unionized mining sector. Carlos was tired of this policy. Once in the COB, he wanted to propose modifications that would strengthen the COB's service to the unsalaried sectors it claimed to represent: "Do we want to strengthen the Bolivian Workers' Central or not? We want to, so let's [do it]. In order to benefit these *cooperativistas*, *transportistas* [taxi, bus, and long-haul drivers], *gremiales* [guilds], *campesinos*, let's make a thesis similar to that of Pulacayo and make it prevail in the Bolivian state" (Uncía, June 3, 2017). The Thesis of Pulacayo to which he referred was *the* landmark document of the Bolivian workers' movement, penned by the Trotskyist historian Guillermo Lora and signed by a group of miners in the town of Pulacayo in 1946. It called for a living wage, forty-hour workweeks, worker control of the mines, arms for workers, and a political vanguard of miners to fight the ruling classes. For Carlos to conjure a similar document that identified the unsalaried workers as the new subject of the Plurinational State was significant in that it refused to cede ground to traditional subjects of history (i.e., unionized workers). He elaborated:

> With this new president [Evo Morales] what's happening is that on one side are appearing some workers, workers that are the spoiled children of the state, and others that are the stepchildren of the state, and which are these people? For example, I can tell you that the spoiled children of the state, those people have raises, those people have eight-hour workdays. . . . Who are these people? They're the factory workers, the unionized miners, also professors, soldiers, police officers. All these sectors only add up to 30 percent of the Bolivian state, imagine. And now the campesino sector, which is around almost five million people; the cooperative sector, some 400,000 people; the freely unionized transportation sector, almost a million. . . .[22] What happens with this other 70 percent? This 70 percent, or let's say some seven million people or so, of the inhabitants of the Plurinational State of Bolivia, these people don't have eight-hour workdays, don't have raises, don't have layoffs or severance packages, don't have . . . I don't know, even in the Ley General de Trabajo [General Law of Work], in the introduction they mention us, they call us "informal workers." (Uncía, June 3, 2017)

What Carlos is describing here might be understood as the precarization of labor in Bolivia. As Bolivian labor scholars have pointed out, the flip side of Evo Morales's strategy of channeling wealth and decision-making to noneconomically defined "social movements" has been a weakened working class. According to the 2012 national household survey, 60.7 percent of workers in Bolivia are unsalaried, falling into a hodgepodge category of self-employed and popular-sector labor (Escóbar de Pabón, Rojas Callejas, and Hurtado Aponte 2016).

But this transformation is about more than just economic restructuring; it is also about the surge of the country's Indigenous majority into urban spaces and middle classes, often while maintaining strong personal and financial connections to rural communities. Indeed, engaged scholars sometimes reframe the "informal workers" that Carlos described as an emerging "Indigenous middle class," a category that is exemplified by the (predominantly female) Aymara merchants who run openly unregulated markets with sourcing networks that stretch around the world (Arbona 2007; Tassi 2017). Until recently, it was not even possible to conceive of an "Indigenous middle class" because indigeneity was conceptually grafted to rural areas, small-scale agriculture, and poverty. Cooperative miners are thus part of a broader rearticulation of Bolivia's racial formation, in which Indigenous peoples can no longer be economically spliced into the *campesinaje* (peasantry).

What does this mean from the perspective of the Plurinational State, which enshrined Indigenous plurality but deepened resource nationalism? How are cooperative miners reconciled with these national projects?

K'EPIRINAS IN THE STATE

On May 27, 2017, President Evo Morales held an event in a stadium in Oruro to honor women cooperative miners across the country. After traveling overnight to arrive from every regional federation, *palliris* and *socias cooperativistas mineras* (women cooperative miners who work underground) gave Evo an enthusiastic welcome that was only partially motivated by the lunch and beer they expected him to provide. As Evo stood to speak, two women wearing helmets and *polleras* approached him and ceremoniously helped him into a *k'epirina* that he wore for the duration of the event (see figure 3.5). The *k'epirina*, the potato sack backpack that exemplifies the agricultural origins of the *agro-minero*, in this case symbolized the mutual support of mining cooperatives and Evo's government.

FIGURE 3.5. Evo Morales wearing a *k'epirina* to speak in front of *socias mineras cooperativistas*. Oruro, May 27, 2017. Photo by author.

Indeed, according to some commentators, the very term *agro-minero* has been encouraged by the MAS (Movement toward Socialism) government to reduce conflict between its varied political bases. As a "social movement state," the MAS had to unite many sectors that did not always see eye to eye, such as Indigenous federations, campesino unions, salaried workers' unions, and collectives of unsalaried workers. Although the term *agro-minero* had existed for decades prior, critics suggest that it was specifically celebrated by the MAS because *agro-minería* diffused antimining sentiment in rural areas while also alleviating rural poverty with very little government effort or investment.[23]

While it's hard to say for sure whether the term was embraced as part of a concerted effort to promote mining, as critics suggest, or if it recognized an already-existing set of livelihood strategies, what is certain is that the mining cooperatives and the MAS government have strengthened one another as the number of *agro-mineros* has grown. As if to mark their alliance, Evo chose Walter Villarroel, a cooperative tin miner from the town of Huanuni, as his first minister of mining and metallurgy in

2006. Villarroel didn't last long—during his first year in office, a conflict broke out in Huanuni between cooperative and salaried miners that left at least seventeen people dead and resulted in Villarroel's dismissal—but his departure by no means marked a definitive break between the government and the cooperative mining sector.[24] Cooperative miners are an influential segment of society, and not one the state can easily ignore.

In addition to acting collectively under the cooperative mining umbrella, *agro-mineros* also slide between and influence other segments of society. In Llallagua-Uncía most cooperative miners are also members of either CONAMAQ, the highland Indigenous federation, or CSUTCB, the national peasant union. Thus, in addition to exerting pressure on the government through the ballot box and as a collective of cooperative miners, *agro-mineros* also influence rural political bodies from within. While it is certainly not the case that all Indigenous or campesino union leaders support mining cooperatives, there is little overt resistance to cooperative mining in the rural highlands; this is in part because the same people occupy multiple political spaces. Bolivians identified with Evo's "Indigenous State" in a variety of ways, but in the highlands this identification was often articulated with a demand to take control of and profit from subterranean spaces that had long been claimed by colonial-nationalist forces.[25]

Understanding how cooperative miners challenge mestizo resource nationalism involves understanding how subterranean tin mines were historically materialized as separate from surrounding (and legally soil-deep) Indigenous territories, and how neoliberalization triggered an increase in two-way traffic between the two spaces. In the early twentieth century, Indigenous men who went to work in the tin mines were imagined not only as becoming members of the proletariat but also as becoming mestizos, potential subjects of a developmentalist economic agenda. When these same miners were drafted into the army during the Chaco War, the discursive convergence of masculinity, *mestizaje*, and nationalism centered around extractive industries, where Bolivians identified the theft of natural inheritances through the collusion of national elites and foreign companies. The prominence of the tin miners' unions in the 1952 revolution sealed an association between tin mining and national popular sentiment, a fact that is widely recognized in Bolivian historiography.[26]

Although most of Llallagua-Uncía's labor force was, in its early years, drawn from the Quechua-speaking valleys, neoliberal policies dissolved the semipermeable membrane that separated enclave mining towns and the Aymara-speaking ayllus of Norte Potosí. Often keeping one foot in the ayllu and one foot in the mine, these *agro-mineros* have transformed the internal operations of cooperative mining in Bolivia. Traditions of inheriting access rights to underground *parajes*, joining extended family members' *cuadrillas*, and rotating the actual labor of mining among *cuadrilla* members all reflect the organization of a family farm. Moreover, kinship networks link the miners working underground to families and land in surrounding areas, networks through which people and resources move with regularity. This does not mean that tin mining was fully dissociated from the nation-building project of *mestizaje*. Instead, these older meanings remain as sedimented layers in contemporary tin mines, where cooperative miners dig them up and repurpose them. This can look like a fragmentation of nationalist sentiment, in which cooperative miners claim their right to mine individually based on shared kinship with the national collective, but it can also look like a shift in primary identification, in which one labors on behalf of the ayllu rather than the union, cooperative, or nation.

More than a rearticulation of livelihood linkages, the emergence of mining cooperatives in the 1980s also represented an "indigenization" of the mining sector and the subterranean itself, as individual *agro-mineros* take both their Indigenous identities and their agrarian practices to work with them underground. Carrying *k'epirinas* and transgressing the surface-subsoil boundary, cooperative miners as *agro-mineros* are shifting what it means to be both Indigenous and Bolivian within the Plurinational State. Whether this is in the long-term benefit of their territories and autonomies, however, remains to be seen.

4

"Here's the vein," Samuel said, pointing his helmet upward to shine light on the ceiling a few inches above our heads. It revealed a barely visible black streak, sandwiched between two lines of white quartz. Samuel and his two other *cuadrilla* (work crew) members had been following the white quartz for weeks, hoping it would lead them to tin, and it finally had. The nearly vertical shaft through which we had just crawled, rubber boots shoved into crumbling footholds, traced their laborious journey. Now they hoped the black streak on the ceiling would not *despintar*— literally, "to lose color," or disappear. Samuel explained that a vein of tin is often more like a pearl necklace than a vein of blood, and you have to keep following the string to collect all the pearls.

We were on our hands and knees in a section of the tunnel that had just been dynamited yesterday, beginning the process of separating the freshly broken rocks into two piles: those that contained tin ore and those that did not. Samuel leaned on my knee and reached around my leg to show me what to look for. He picked up a rock with a thick sheet of black on one side. "This is a good one. Feel how heavy it is—it's at least 55 or 60 percent tin," he said, dropping it into my hand. He leaned back and looked around his own knees for another rock for comparison. "This one is less than 30 percent tin—it's *q'ara*," he said, chuckling a little. *Q'ara* is the Quechua word for "peeled," but it is also a term for non-Indigenous Bolivians, whose paler skin and proximity to foreign powers have left them culturally "peeled." In the *q'ara* rock, the black streak was lighter, with an almost orange-red hue. It contained a tin sulfide that makes the water inside the mine run orange. "Anything lower grade than that isn't worth exploiting," Samuel added.

This chapter centers moments of encounter and transformation between workers and rock. Cooperative miners transform tin ore into a commodity through activities that range from dynamiting to hammering, drilling, washing, and sorting. Yet it is not just minerals that are transformed through these processes; miners are also constituted and hierarchically ordered along raced and gendered lines through their specific material intra-actions with the rocks, their mediating tools, and the spaces within which they work. Borrowed from Karen Barad (2007), the term *intra-action* signals how differentiated matters are always coming into being in relation to one another, rather than "interacting" as ontologically discrete objects. Rock, water, skin, nails, clothes, tools, flesh, and a host of other material things are packed together in cramped underground spaces, their intra-actions permanently changing both human bodies and ore bodies.

Throughout this chapter my methodological assumption is that sites of labor are significant sites of (human) subject formation. In this, I both draw on and depart from Marxian historical materialist accounts of political consciousness. Like historical materialists, I am interested in precisely how humans are formed and unformed through their efforts to change the matters of nature into commodities. The sensuous qualities of nature matter, as do the habitual movements of the body in its habitual spaces. My focus, however, is on the constitution of social differences through laboral intra-action, a focus that is inspired by feminist science and technology studies. As geographer Kathryn Yusoff (2018, 202) elucidates, there is and has long been conceptual traffic between the classificatory systems of geology and the classificatory systems of race. In her analysis the notion of the "inhuman," or that which passively *has* properties rather than *owning* property, is "a connective hinge" between geological economies and racial hierarchies. Anthropologist Elizabeth Povinelli

(2016) also argues that geology and its corresponding economic exploits (prospecting, mapping, engineering, smelting, etc.) all depend on a language of materiality in which life and nonlife are neatly distinguishable; in colonial encounters, this organizational system also enabled the distinction between human and nonhuman. Extending this analysis, geology's internal taxonomies, in which rocks are ordered according to their metallic content and corresponding economic value, correspond with racist orders, in which people are similarly classified and hierarchically ordered.

Thinking with Barad, Yusoff, Povinelli, and Karl Marx, I contend that the site of labor is not only a site of individual formation but also ground zero for hierarchically ordering people and rocks along related axes of value. These axes of value are racial, as Yusoff and Povinelli have detailed, but I argue that they are also fundamentally gendered. As transnational and Black feminists have been arguing since the late 1970s, the two categories of difference grind against one another in a way that transforms experiences of each.[1] They have collectively shown that it is impossible to adequately attend to questions of gender without considering race, and vice versa. With this in mind, my argument proceeds at three levels.

First, I continue to develop my argument that all matters, whether mineralized or fleshy, are always already historied before becoming enmeshed with one another at the site of labor. These material histories are both biophysical and sociocultural. As I explored in chapter 2, tin compounds were crystallized by forces beyond human spheres of influence (tectonic plates, magma plumes) long before humans were around to identify them as such, but tin did not exist as a "pure" metal in nature: there is quite literally no metallic tin as we know it outside of human processing techniques. As Barad reminds her readers, moreover, meanings and matters are inseparable, and the processes by which humans have materialized tin as a standalone element are richly discursive. In this case, there is no tin outside of the meanings that have been attributed to it. In Bolivia, as argued in chapter 3, these meanings have been constituted by the promises of economic progress and entry into a national community dominated by masculine and mestizo identity politics. Tin comes into Llallagua dripping not only with blood and dirt, as Marx argued about capital in general, but also with expectations of wealth and progress.[2]

Second, I argue that tin's mineralogical variation—both that which occurs "naturally" and that which has been produced by a century's worth of extraction—crystallizes social stratifications among miners. Put differently, racialized and gendered hierarchies within mining cooperatives are constituted in relation to the material specificities of tin as an element and as a geological formation. Tin is contained in a variety of ores with a variety of grades, and these ores are both unevenly distributed across volumetric space and differentially valued in economic and sociocultural terms. Individual rocks, veins, and whole sections of the mine are valued according to their metallic grade, and this hierarchy of value corresponds with racialized and gendered differences as they manifest within and around the mine. I am not just arguing that women and Indigenous people have access to poorer-quality ore, although this is also true; I am also arguing that the material axis of value through which rocky matters are classified is mutually constitutive with the raced and gendered axes of value through which miners are classified. For instance, when rock dust settles in miners' lungs, or when miners identify one another by the metallic smell that adheres to their skin, the meanings associated with these matters literally and figuratively constitute the miners' bodies, marking them as someone who drills (and inhales more dust) or works underground (and smells more like the subterranean).

Finally, this chapter explores how geological and social axes of value change over time in connection with one another. Minerals and miners are relationally valued in ways that shift not only spatially but also temporally. To get at the temporal aspect as well as the spatial, I use the concepts of *formation* and *degradation* to act as connective tissues between geological and fleshy matters. Both words can be used to describe social and environmental processes, but they work against one another: *formation* suggests a coming together over time, while *degradation* suggests a slow coming apart.[3] As discussed in preceding chapters, cooperative miners now are very different from the unionized miners of decades past; no less important, the mine is a very different place now than it used to be. The more tin that miners extract, the lower the average ore grade. Time does not stop just because capital takes flight, and miners and the mine continue to move through one another.

DEGRADATION

The word miners most frequently used to describe Llallagua's Juan del Valle Mountain is *agotada*: exhausted, worn out, used up. Occasionally, they also used the word to describe themselves, or more specifically their lungs, which had weakened after years working belowground. In that it is used to describe a shared condition that the mountain and the miners bring on one another, *exhaustion* is an interesting word to sit with.

Samuel, the miner with whom I opened this chapter, used the word before we even entered the mine. We were sitting outside the Siglo XX mine shaft, eating a breakfast of lentil soup purchased from a young woman who always showed up at the crack of dawn to feed the miners. Perched on a precarious-looking slab of rusted metal, we watched as Hilario, an elderly miner who guarded the entrance of the mine and repaired shared equipment, hooked together a collection of ten or twelve iron carts to form the trolley that functioned as the miners' commuter train into the mountain. Hilario was struggling to make it work, and Samuel gestured toward him with his chin. "The mine is exhausted," he said, "and so is our trolley. It's been broken for days. We might have to walk."

An argument that runs through Povinelli's (2011) work on Australian settler colonialism is that endurance and exhaustion are differentially distributed across social difference: while some people may be enduring/surviving/adapting to increasingly constrained ways of life under late liberalism, they are unevenly equipped to resist a slide toward exhaustion/collapse/surrender. In much of her analysis, the line between the two conditions seems to resemble a cliff. One can be on the brink of exhaustion for a long time, but it is hard to climb back once pushed over the edge. In her later work on geontology and geontopower, Povinelli (2016) expands her discussion beyond living worlds, arguing that the governance of difference and the uneven distribution of harm is itself predicated on an ongoing separation and ordering of life and nonlife. Like Yusoff, she argues that this ordering makes it possible to justify plumbing the depths of the earth to sustain (human, capitalist) life, while simultaneously making it possible to order people based on their ability to distinguish subjects from objects, or to appropriately assert their will through the transformation of these objects. The grammar of life and nonlife, rather than life and death, enables both capitalist resource extraction and colonial-racist objectification.

In this chapter I use the word *degradation* to signal something similar to the slide from endurance to exhaustion that Povinelli describes, with the emphasis on the process of deterioration rather than the contrast between the two states. The multiplicity of shades assumed in the word *gradation* draws attention to variations in the experience of the slide. Degradation operates across the life/nonlife divide, though the process signaled varies depending on what is being degraded. In ecology a degraded environment is one that is damaged, contaminated, or otherwise less supportive of life. In resource economics degradation signals the loss of raw material and a correspondingly lower level of productivity; for nonrenewable resources, degradation is inherent in the process of exploitation. In a social context, degradation is typically used to describe the construction or conjuring of social hierarchies in which one person (or group) degrades another. Like waves, all these forms of degradation—environmental, economic, and social—amplify each other. In Llallagua-Uncía rocks and people are graded and degraded together, but it is important to note that there is no cliff here. Degradation is asymptotic: there is no point at which it is complete, just an endless series of increasingly imperceptible gradations of difference.

When Samuel told me that we might have to walk into the mine because the trolley was exhausted, I was not thrilled. Since it was the rainy season, and since we were entering from a lower level of the mountain (650 meters down from the peak), it would be a tiring two-mile trek through either thick mud or standing water. I was therefore relieved when Hilario seemed to resolve the trolley issue. He heaved himself into the front conductor's cart with some effort, using his hands to assist one of his legs, which had been crushed in an underground accident several decades beforehand. About a hundred miners piled into the trolley carts after him, with Samuel and me among them. We took off with a lurch, with Hilario operating the trolley by touching a long copper rod to the exposed electrical cable that ran along the main passageway of the mine. As the trolley entered the shaft, the ceiling dropped so low that we had to lean way back to avoid being decapitated. The wooden beams that act as the mine's exoskeleton are over a century old, and they buckle and sag in places that are familiar to the miners, who occasionally shouted "Knees!" and "Heads!" as a collective reminder. Otherwise, no one spoke—it was far too loud, with the rusted wheels screeching along the tracks and the crackling sound of electrical current in the copper rod.

Being underground is a full-body experience. Nothing is visible beyond the circular beam of your headlamp, but all your other senses are activated. Depending on what section of the mine you are in, the air is either cold and dry or hot and humid, but no matter what, there never seems to be quite enough oxygen to go around. Everything is wet. You slop through pond-size puddles of acidic orange water known as *copajira*, and the noise of your boots is deafening in the otherwise thick silence of the mine shaft. The walls themselves seem to sweat and are covered in spongy growths that, to my inexpert eyes, could be molds or lichens as easily as mineral deposits. Miners have left so much biological detritus over the years—including coca leaves, bean shells, cracker crumbs, and toilet paper—that I imagine molds would have no trouble spreading. In between the spongy rocks are nonbiological additions: heaps of soda bottles, beer cans, plastic bags, dropped gloves, broken tools, and colorful decorations used to bless the mine during ceremonial events. Electrical wires crisscross the ceilings of the main levels, often dangling perilously close to your head. Plastic tubing loops down almost all the passageways, some used for pumping out water, others for pumping in air. The smell that encapsulates and communicates all of this—a smell that resists being rinsed from your skin, hair, or clothes—is metallic and musty. The mountain, which seems from the outside like a perfectly obvious geological *thing*, is internally riddled with life, death, and nonlife, with geological, biological, and distinctly social elements folding in on one another in the darkness.

The entrance to Samuel's *paraje* (work area) is a vertical tunnel covered with a cement lid, much like a maintenance hole cover in a city street. To get inside, we shimmied down with our feet on the wall in front of us, our backs against the wall behind us. Ten feet down, the tunnel went horizontal again, and we began scrambling on hands and knees. I had to pull myself forward at an angle because my hips could not otherwise fit, and neither of us could turn our heads sideways to talk because our helmets were too wide. All other bodily sensations aside, the size of these tunnels is what most defines miners' experience underground. Although the old shafts, which serve as the mine's thoroughfares, are spacious enough to stand upright and walk three abreast, this is not the case for the miners' *parajes*. Cooperative miners earn money only when they are extracting mineral, so they do not waste time building passageways any larger than is strictly necessary.

As Samuel and I crawled through such a minimalist tunnel, I felt an explosion of dynamite somewhere nearby, marked in the enclosed space by a bodily popping feeling rather than a noise. Dislodged pebbles sprinkled harmlessly on our helmets. I glanced up and saw a couple logs of wood—surely scavenged from prospector Simón I. Patiño's original infrastructure—jammed into the ceiling to prevent it from caving in. Damp splinters of wood also fell from the beams along with the pebbles, contributing to the mixture of organic and inorganic matter beneath our hands and knees. "You exhausted already?" called Samuel when I paused and panted for breath. "We still have a lot farther to go!"

In discussing exhaustion, Povinelli's (2016) work dwells in places left abandoned by capitalism and late liberalism, and she pays attention to the "alternative social projects" that emerge in these places. For example, she shows how the very inhospitality of toxic wastelands to human life makes possible specific exercises of Indigenous sovereignty in northern Australia (90–91). This is also the case for the Juan del Valle Mountain. Mining cooperatives occupied the mountain only in the moment of its abandonment by the state corporation, which, when it closed in 1985, was working against not only the lowest tin prices in a century but also rapidly declining tin grades. In the early 1900s, shortly after Patiño laid claim to the mine, the tin grade was 17 percent; when the mine was nationalized in 1952, that number had fallen to 1.1 percent; and by the time the state corporation closed its doors in 1985, it was well below 1 percent (Espinoza Morales 2013, 51). This was the death knell for the unions, but it was also the condition of possibility for the cooperatives. Too degraded for the state to profit from its exploitation, the mine was available for occupation by both former miners and migrants from the surrounding rural areas (see chapters 2 and 3).

Yet this same condition of possibility—the degradation of the mine— foretold increasing financial precarity and bodily harm for the miners. Cooperative miners are still extracting tin, which means that degradation of the mine continues apace. Mountain and miners are not only mutually constitutive, as geographers argue about nature and society in general, but bound in a mutually harmful embrace.[4] This is why, in addition to social forms of degradation, I also think about bodily degradation, understood as the fleshy harms experienced in intra-action with particular (toxic, dangerous) environments. Accidents are all too common underground. Harmful gases can build up in enclosed spaces and kill whole work crews in an instant. Century-old elevator cables can snap. Rotting ladders can break. A miscalculated stick of dynamite can bring down the ceiling. Even a tiny stone falling from a great height can take out an eyeball. The experience of working in these subterranean spaces shows up on and in all the miners' bodies in the form of scars, injuries, reddened eyes, burning lungs, aching backs, and calloused hands.

But degradation is complicated here, for social degradation does not always align with bodily degradation. Although subterranean work is more dangerous than surface-level work, few subterranean miners would ever choose to transition to surface-level work. Positive meanings adhere in the bodily transformations that come from working underground, which function as fleshy evidence of self-sacrificial bravery—traits associated with masculinity and mestizo nationalism. When two substances come into contact, the meanings attributed to their intra-actions depend on their individual historical formations; and things of the mine, even the harmful things, have been historically formed in ways that turn miners' bodily injuries into badges of honor. Thus, the three forms of degradation—of the mine, of the individual body, and of the social body—inform one another across volumetric space.

FORMATION

Formación, or formation, is the word used throughout Bolivia to discuss how individual people have come to be who they are. It is also used, more pointedly, to assess someone's political commitments. Wherever I went, but especially in Llallagua, people would ask me, "Qué es tu formación?" Translated, this question can be understood with equal accuracy as "What have you studied?" and "How have you been shaped as a person?" But what people really wanted to know was what I stood for and—in the case of a conflict—who I would stand with. The question was often followed by specific inquiries about books I had read or activism I had been involved in.

The importance attributed to someone's formation has a history that extends far beyond the need to evaluate resident gringas. Up until 1985, when Llallagua was still a union town, miners believed that workers' individual formations would determine whether they stood for or against the collective spirit of trade unionism. At this point in history, tin miners prided themselves on being Marxist and Trotskyist autodidacts, and they took mutual education very seriously.[5] To serve their children and institutionalize a trade unionist spirit, miners in Llallagua-Uncía spent more than two decades, from the 1960s to 1980s, fighting for the construction of a local "workers' university." Tragically, they won this battle the same year that neoliberal restructuring tore their unions apart. But their spirit lives on in the university, which was named—after the mine itself—UNSXX (Universidad Nacional Siglo XX; National University "Siglo XX" [Twentieth Century]).

Although it now serves college students from across the country, most of whom are studying dentistry, nursing, or engineering and have little or no connection with miners' unions, the defining feature of UNSXX is the existence of an FPS (political syndicalist formation) department. All students are required to take a series of three classes in this department, which has the explicit aim of producing "organic professionals" who will serve the working class even while climbing social ladders. I attended a handful of FPS classes in which this purpose was explicitly addressed. "What is an organic intellectual?" one of the FPS professors asked his students. "It is he who is always in service of his community. . . . Our consciousness/conscience must be committed.[6] We are the extract of the class of campesinos, of miners, and when we get ahead, are we going to turn our backs on our roots? No, we should not, and we will not."

This unionist interpretation of *formación*, however, does not live on in the same way among cooperative miners. The week after I attended the FPS class, I recounted the experience to Don Esteban, a middle-aged cooperative leader who had become my unofficial advocate among other miners (he will return in chapter 6). He laughed: "Good! You will have your political formation over there with them and your counterformation underground with us." We were sitting in Don Alfonso's office in the FERECOMINORPO (Regional Federation of Mining Cooperatives of Northern Potosí) building in Llallagua, and the other five miners present, including Alfonso, joined in the laughter. They seemed to approve of Esteban's assessment of the FPS department, that it was out of touch with the reality of the mines.

It was hard to disagree with Esteban. Although only one member of the FPS faculty had himself been a miner, all of them idealized the unions and spoke disparagingly of the cooperatives. One of them had even moved out of Llallagua to the nearby town of Huanuni, where the last remaining state-owned tin mine was still in operation, so that he could live among unionized workers. The single faculty member who had been a miner himself tried to give seminars to the cooperatives, but he was not enthusiastically received. When he showed up with his slides about global imperialism, the *cooperativistas'* faces glazed over. They stayed for the snacks but not for the world systems theory.

The word *formation* is useful for thinking across the different ways that the FPS faculty and cooperative miners were conceptualizing subjectivity. *Formation* signals coming together, and the word is used in both geology and social theory. Rock formations are historical products: they have taken their current forms only after millennia of sedimentation, erosion, volcanic intrusions, pressure, and heat, and they are continuously changing through the same processes. The aggregate of minerals that define a rock formation could only have emerged through the precise combination of forces that characterized its history. Although formations with similar histories from different regions of the world are classified together, no two are identical. This is also true for social formations, both individual and collective. As Antonio Gramsci argued, the *persona* (person) is a historical product, having taken their current form as a result of both their individual experiences and the sedimented histories that define the social context through which they move (Thomas 2009).[7] Similarly, Michael Omi and Howard Winant's ([1986] 2015) influential theory of racial formation takes history as fundamental to the emergence of both broad racial structures and individual experiences of race and racism. People are historical products, just like rocks.

In contrast, the faculty members in the FPS department seemed to think primarily about current economic rather than historical structures. They repeatedly told me—often as they tried to convince me to pursue an alternative research topic—that cooperative miners are not laborers in a Marxist sense, and this is why Marxism does not resonate with them. Rather than trading labor time for wages, cooperative miners hold their means of production, which makes them less like genuine workers and more like petit bourgeois capitalists or mutated peasants, depending on who is doing the categorization. The FPS professors were working primarily with social categories lifted from the pages of Marx's *Capital* ([1867] 1990), often as interpreted by local Trotskyist historian Guillermo Lora. Like pre-neoliberal union leaders, *loristas* believe that political consciousness grows out of an awareness of collective, structural exploitation.

Cooperative miners could not have disagreed more. Instead of insisting that I understand their structural position within an economic field, Esteban and the other *cooperativistas* wanted me to understand their labor. They wanted me to know how far they walked with their backs doubled over, to recognize the jolt of fear that tingles in their extremities with every blast of dynamite, to feel the dull ache of their lungs after a day spent drilling in enclosed spaces. They agreed that labor mattered but thought that it mattered differently.

To take seriously both the structural principles offered by the FPS professors and the experiential labor processes emphasized by the cooperative miners, I found myself returning to Marx's early writings on labor and consciousness. In their notes on a materialist theory of history in *The German Ideology* ([1846] 1998), Marx and Friedrich Engels begin with a detailed analysis of individuals in relation to their transformations of nature. The following quote contains two sentences in brackets that Marx and Engels wrote and then struck out, which I have included because they instructively point to the theoretical lineage within which the authors were working:

> The first premise of all human history is, of course, the existence of living human individuals. [The first *historical* act of these individuals distinguishing them from animals is not that they think, but they begin to *produce their own means of subsistence.*] Thus the first fact to be established is the physical organization of these individuals and their consequent relation to the rest of nature. Of course, we cannot here go either into the actual physical nature of man, or into the natural conditions in which man finds himself—geological, oro-hydrographical, climatic and so on. [These conditions determine not only the original, spontaneous organization of men, especially racial differences, but also the entire further development, or lack of development, of men up to the present time.] All historical writing must set out from these natural bases and their modification in the course of history through the action of men. (Marx and Engels [1846] 1998, 37)

Marx's "materialist method" thus begins not with a set of economic relations but rather with the relationship between "men" and nature. In this way, his method was not as radical a departure from that of his theoretical predecessor, Georg Wilhelm Friedrich Hegel, as Marx would have us believe. Among other hats that he wore, Hegel was a geographer of sorts, and he wrote extensively about the determining impact of environment on (racialized) human development. Through transforming their surrounding external nature, Hegel argued, "men" move out of their own internal "state of nature" and attain higher planes of self-consciousness (Hegel [1822–28] 1997). While dropping the overt environmental determinism and literally crossing out the sentence that connected nature to race, Marx retained Hegel's emphasis on the transformation of sensuous nature through labor as key to the development of human consciousness.

The racializing implications of Marx's reading of consciousness and labor might seem like a good reason to avoid Marxism, or at least to pry his theories of individual formation away from his theories of class formation. But I want to try the opposite. By holding on to both Marx and critiques of Marx, labor becomes a site productive not only of class consciousness but also of racial and gendered formations. The process of engaging the matters of nature with transformative intent—especially when the matters of nature, and all the tools required to transform it, are understood to have been discursively constructed prior to this engagement—contributes to the ordering of people along hierarchies of racialized and gendered value. The point is not that nature is determinative but rather that human bodies are materialized and valued in the same moments that nature is materialized and valued by humans. Labor thus matters for subject formation but not *just* in class-based ways.

To take this point seriously, matter itself requires more attention. Marxian historical materialism, although it certainly begins with the sensuous qualities of nature, tends to emphasize the importance of material relations over material stuff. Quite the opposite is true of new materialities, a body of literature that tends to emphasize the agential or lively qualities of matter rather than its integration into a social world structured by relations of power. In a new materialist analysis, nonliving matter and nonhuman creatures enter into convivial assemblages with humans, generating specific effects and affects that are not wholly the work of any single agent/actant. Scallops, fishers, and scientists end up entangled around mollusk population collapses (Callon 1984), financial interests mix with electrical grids to cause enormous blackouts (Bennett 2010), and mosquitoes interact with networks of canals to transform imperial relations (Mitchell 2002). Labor is not a central site of analysis in the majority of these texts—nor, frankly, are people *as* people, with their variegated individual and collective histories. Although the new materialist literature often provides fascinating analyses of unlikely alliances, the normative edge is hard to locate in much of it. Between these two extremes, an apparent middle ground is occupied by political ecologists who are primarily interested in evaluating how the specific physical and chemical qualities of nature disrupt or permit human efforts to engineer the world in a way that maximizes specific outcomes, whether those outcomes are economic, political, or social. Water flows, fish swim, viruses mutate; nature is difficult to control, no matter how powerful the social world appears.[8] Usually cleaving closer to the Marxist tradition than new materialism, however, these political ecologists tend to zero in on single, economically important resources (sugar, cows, water, oil, turfgrass, etc.), downplaying substances or sentiments that don't ultimately figure into economic processes.

This is where Barad's work on materiality is fundamental. Barad argues that not only are matter and meaning inseparable, but the former comes into existence only through the latter. Drawing inspiration from Judith Butler's work on sex (an apparently material attribute), gender (an apparently meaningful attribute), and the way the two come into being together through performative practices, Barad insists that performativity is not limited to people. All matter comes into being performatively, in actions that are as deeply meaningful as they are quotidian. In this ontological spirit, other scholars draw on Barad to show how matter is not only meaningful but also meaningful in ways that are specifically unjust. They demonstrate how, for example, racial difference is mattered in lead-painted toys imported from China into the United States (Chen 2012), harmful agricultural toxins are inextricably linked to both nationalism and homophobia (Agard-Jones 2013), and polluted landscapes can become sites for reclaiming sovereignty (Murphy 2017; Povinelli 2016). Infinitesimally small particles contribute to the performance of racialized, gendered, and sexualized bodies while also threatening socioenvironmental destruction on massive scales. These are the new materialities that inspire this chapter.

Integrating all these insights into the labor process means insisting that both rock and flesh have histories that are historically antecedent to their encounters with one another and that these meanings shift— along with material transformations—through the process of labor. In what follows, I use stories from miners' work sites to continue to show how material things and material spaces shape individual bodies as well as social formations, as divided along raced and gendered lines.

ROCKS OUT OF PLACE

Before we went to his work area, we stopped at Samuel's *pauwiche* to wait for the rest of his work crew. A *pauwiche* is an underground storage room, but it is also the site of the requisite morning *ch'alla* (blessing), which involves chewing coca and drinking alcohol, part of which is spilled on the floor as an offering to Pachamama (Earth Mother). Samuel's *pauwiche* was locked when we arrived, its sheet metal door held shut with a padlock. Despite this security system, someone had managed to break in and steal the lightbulbs the week before, so we had to sit in the dark—save the light from a single headlamp—while we chewed coca and talked.

As we waited for Tomás and Rosa, the two other members of his *cua-*

drilla, Samuel recounted how he had become a miner. He had migrated to Llallagua from the ayllu Aymaya in 1995, when he thought he could earn more mining than farming. He had learned both Aymara and Quechua at home and Spanish at school; he described his parents as *originarios* (original or Indigenous people). In Llallagua he began working as day laborer, milling and concentrating tin for other miners in exchange for a daily wage, before amassing enough money to become a full member of the cooperative. Then he began working underground in a *cuadrilla* led by Marcos, a former union leader who had used his influence to lay claim to one of the richest sections of the mining cooperative Siglo XX's concession. Marcos, with whom I later became acquainted, was a third-generation tin miner whose grandfather had migrated from the Cochabamba valleys to work in Patiño's mine prior to the 1952 National Revolution. Although he also spoke Quechua, Marcos had grown up in Llallagua speaking Spanish with everyone other than his mother, and he identified as mestizo.

In the *pauwiche* immediately next door to Samuel's, all the lights were on, and we could hear a large group of men laughing and talking. Quietly, so they wouldn't hear us, Samuel told me that these men were campesinos from the ayllu Chullpa, which—as described in chapter 3—had occupied the mine back in 2005. At the time, they had claimed that they had a right to the mine because it was located on their ancestral territory. They had been particularly interested in the region where Samuel was working with Marcos's *cuadrilla*, as rumor had reached them of Marcos's lucky strike. To settle the violent conflict that emerged, the cooperative Siglo XX agreed to accept a hundred Chullpa members into its ranks and granted them access to the rich section that Marcos's *cuadrilla* had been exploiting. The men next door, Samuel whispered, were the ones who had instigated the conflict and who continued to profit from veins that had once belonged to Marcos.

Samuel's most repeated complaint about the neighbors, however, was not that they had started a violent conflict and stolen his *cuadrilla*'s territory, though that was important background. Instead, he said over and over how frustrated he was that they had made the mine so dirty. Although anthropologist Mary Douglas has influentially argued that dirt is a culturally specific classification that broadly signals "matter out of place," it struck me as a particularly hard category to interpret underground.[9] The walls oozed continuously, the floors were littered with ground-down coca leaf stems, and the ceiling seemed to threaten to become the floor at any given second. Thinking that he was perhaps talking about the piles of plastic water bottles and snack bags—my own understanding of dirt—I asked Samuel to clarify what he meant. Plastic, it turned out, was not the issue; the issue was abandoned waste rock, or *caja*. *Caja* is the Spanish word for "box," and *caja* rocks are just that: empty boxes, containing nothing of value. Samuel spoke wistfully of the days when Marcos had been in charge, when he and his *cuadrilla* had spent whole days diligently clearing the passageways and galleries of *caja*. "Then the floors were *limpiacitos* [very clean], no *caja* anywhere," he mourned. "Not like now, with all these campesinos." He was not the only one to make such comments to me; on a later occasion, Hilario, the retired miner who drives the morning trolley, toured me around the mine while complaining continuously about how dirty (i.e., full of *caja*) it had become since the cooperatives had started admitting migrants from the ayllus.

From one angle, it's practical to clean up *caja*, since it is much easier and safer to maneuver in a space with fewer loose rocks, but from another angle, it is an enormous waste of time and bodily energy. In the unionized era, the company paid its workers an hourly wage to clean up the *caja*; in the era of the cooperatives, when miners are earning based on quantity of ore rather than hours worked, few have leftover strength to move worthless rocks out of the way. *Caja* may not contain any tin, but it is still heavy, and clearing it is not particularly "rational" from an economic perspective. Samuel's and Hilario's critiques were more about distinguishing between right (orderly/clean) and wrong (disorderly/dirty) forms of mining—and right and wrong kinds of miners. In Samuel's case, he seemed to be drawing a distinction between the practices of the miners he described as *campesinos*—who would have likely called themselves *originarios*—and his own knowledge of how to "properly" care for and clean the mine. My guess is that he was concerned I might miscategorize him as one of them, given that he was also from a family of *originarios*. He had learned from ex-unionized miners how to maintain subterranean order, a personal experience through which he identified more with multigenerational mestizo miners like Marcos than with other recent migrants from the ayllus.

The distinction between dirtiness and cleanliness often masks a racial distinction, in Bolivia as elsewhere. Throughout most of the world, the colonial normalization of the white, cis-heterosexual, male body was made possible by the simultaneous naturalization of dirt and disease in colonized bodies. For instance, Anne McClintock (1995) discusses how nineteenth-century British soap advertisements used drawings of Black and brown bodies having their color "washed" away, a depiction that both encouraged the cult of domesticity in Victorian England and justified various "cleansings" abroad, while Saidiya Hartman (2019) exposes how early twentieth-century social reformers in New York blamed the interior spaces of Black-occupied tenements for producing wayward communities. In the Andes Benjamin Orlove (1998) argues that middle- and upper-class Peruvians understand the presence of earth—whether dirt floors, earthenware dishes, or unshod feet—as a marker of "Indian" identity, and Rudi Colloredo-Mansfield (1998, 186) similarly notes that in Ecuador "white-mestizos still use pernicious images of disease, irrationality, and 'dirty Indians' to characterise *indígenas* and justify their poverty."[10]

In all these examples, spaces perceived as dirty or disorderly were treated as if they might infect, by proxy, those who regularly occupied them. This is also true of the Juan del Valle Mountain. Not only is *caja*, as dirt, excluded from the taxonomy of valuable rocks, but its ubiquitous presence degrades the mine, which is no longer as orderly as it used to be. Moreover, the miners who fail to clean *caja* away, either because they do not know that it is dirt or because they do not care, are collectively degraded by their supposed neglect. Their association with *caja*, rocks whose very name speaks of worthlessness, was invoked by Samuel and others to imply that they had been improperly formed as workers and were therefore less worthy of access to the mountain's interior spaces. The distinction Samuel drew was racial, but it was materialized through a distinction between rocks that belonged and those that did not. What mattered was who could tell the difference.

CLOTHES AND CALLUSES

Samuel worked with two other miners: his son Tomás, in his twenties, and his cousin Rosa, in her fifties. By the time Samuel began telling me about the campesinos next door, Rosa, who is one of only ten women who work underground in this mountain, had joined us in the *pauwiche*. She nodded her head sagely throughout Samuel's discussion, in full agreement with his assessment. A few moments later, when the supposed campesinos tried to start a conversation through the wall, she yelled at them to "grow some balls" and "knock on the door like men" if they wanted to talk. Coming from a woman, these jeers must have been particularly insulting, and the miners next door quieted immediately.

Rosa's taunts also demonstrated how the practices of racialization and feminization intra-act, which struck me as ironic because of how she herself transgresses racial and gendered boundaries. In her aboveground life, Rosa dresses *de polleras* (literally, "of skirts"), meaning that she wears the clothing of urban Indigenous women, also known as *cholitas*: pleated skirts, sandals, straw hats, blouses, and *mantas* (shawls) pinned around the shoulders. Underground, she wears men's work clothing and is often indistinguishable from her male companions. She jokes that she "transforms herself" every morning when she changes clothes. When I met her, she had been doing this daily transformation for eight years, since her husband left to work in a tungsten mine near Cochabamba and never returned. At first, she worked in her husband's abandoned *paraje*, but she joined her cousin's *cuadrilla* when he found a richer vein and invited her to help with excavation.

Rosa's presence was unusual, as the subterranean is a highly masculinized space. The vast majority of subterranean miners are men, and women are mostly restricted to the labor of *palliris*, who sort through rocks left discarded on the surface in search of remnant traces of ore. This gendered division of space has a very long history that is rooted in colonial labor policies. In pre-Incan communities, labor was organized by the principle of complementarity, which expressed the interdependence of men's and women's tasks; during the Incan Empire, this principle extended to mines, where women and men were sent in pairs (Silverblatt 1987). After the Spanish conquest in the sixteenth century, only men were officially subject to the conscripted labor of the *mita*, but women and children often ended up accompanying partners to the mine, where they tended to small crops and livestock as well as sorting and transporting ores (Absi 2006b). This is the context in which *palliris* emerged, performing surface-level work that may have been complementary to the subterranean work but was not valued as its equal. By the time Patiño was recruiting workers for his tin mine, *palliris* were fixtures of the Bolivian mining landscape.

When I was working in the COMIBOL Archives in Catavi, I located and reviewed the files of 310 women hired by Patiño between 1927 and 1929. Eleven young women from the cities of Oruro and Potosí, who wore their hair in flapper-style bobs cut well above the pressed collars of their blouses, were hired as telephone operators; the others were all hired as *palliris*. These 299 women stared grimly at their photographers, their hair bound in two windswept braids and their *mantas* held tight across their chests. The majority were either unmarried (ages sixteen to twenty-five) or widowed (ages twenty to fifty-one), although a few were married to living miners. All of them started earning between 1.5 and 2 bolivianos per day, compared to between 4 and 4.5 bolivianos for men who began working in the same years. Between 1933 and 1936, ten of the *palliris* began working underground in the context of labor shortages caused by the Chaco War, but their wages only rose to 2.4 bolivianos per day; by 1940 all ten had been fired from their underground positions, with the reasons cited as *falla* (failure to show up) or *flojera* (laziness).[11] Nevertheless, women were not *officially* banned from the subterranean until after the 1952 National Revolution (Absi 2006b). Only at that point was the gendered distinction between surface labor and subterranean labor inscribed in company policy, even if it had been standard practice long before.

But it is not just that only men worked underground; men also *became* men through their subterranean labors. Like joining the army, working underground conferred masculine virtues such as bravery and self-sacrifice, and subterranean miners had a reputation for being willing to take risks. Historically, this risk-taking behavior was assumed to transfer from their daily labors into political activism, and underground miners were considered more politically conscious, or "well formed," than their aboveground counterparts. All the great leaders of the unionized era, those whose statues line the main plaza—Federico Escobar, Isaac Camacho, César Lora—were underground miners. Although cooperative miners are less invested in cultivating political consciousness, they still value the underground for its ability to form fighters, an association that influences decisions about who gets to work underground and how they are treated. The space is also graced with a particularly heterosexually masculine decor: pornographic images of (mostly white) women are everywhere, nailed into dirt walls and pasted onto the sides of large machinery. Although it is relatively normal to find images of scantily clad women in male-dominated workplaces throughout Bolivia—I have seen them floating across screensavers in government and university offices—the ladies of the mine are particularly lascivious, rarely retaining even the smallest scrap of clothing.

A handful of women now work underground, including Rosa. This might be interpreted as the beginning of the end of subterranean patriarchy, but I am not convinced. The women who work belowground are called *socias cooperativistas mineras* rather than *socias cooperativistas palliris*, and they were always very concerned that I not mischaracterize them as *palliris*. The urgency of the distinction emphasized to me the ongoing gendering of the two realms, the aboveground and belowground, as well as the fragility of the victory the *mineras* had won with their upgraded title. For their part, *palliris* whisper that *mineras* have become more like men than women, even making crude jokes like men. It is true that many of them enjoy a lewd sense of humor. For instance, when I was underground with only Rosa and her friend Demetria, I took the opportunity to ask them about the pornographic images on the walls. "Oh, those women are a comfort for the men," Demetria dismissed me, directing her response to the buxom blonde who was watching over us from on high. I asked what comforts the women miners had in the underground, and she smirked: "Well, don't you have a boyfriend? Maybe we can take some photos of him when he comes to visit." All three of us laughed at the idea of pinning up nude gringos. But the point remains that the underground is masculinizing even when occupied by an increasing number of women.

At the time when I was underground with Samuel and Rosa, she was the delegate for women in her cooperative. Between coca leaves, I asked her if she aspired to represent women miners at the regional level in the upcoming leadership change. She shook her head no and then began to complain about the recent leadership transition that had taken place at the national level, in FENCOMIN (the national organization of mining cooperatives). The woman who had been voted in as the representative of woman cooperative miners nationwide, Rosa said, did not know *real* work, she only knew office work—how could *she* represent anyone? I had been present at this election and remembered being surprised myself by the woman who had won. From the city of Potosí, this woman was *de vestido*, meaning that she wore Western women's clothes—in this case, jeans, high heels, and dangling earrings, all of which was complemented by a fresh manicure. She had been running against two other women, one who was *de polleras* and one who wore mechanic's overalls, both of whom held up their callused hands to demonstrate their knowledge of mining.

Among miners, hands are often used to evaluate whether someone's claims to have worked hard are to be trusted. Hands have been referenced in nearly every leadership debate I've observed, the vast majority of which involved only men. More often than not, those who won had proudly displayed their cracked fingernails, calloused palms, and fingerprints whorled with sedimented earth. Their knowledge of the mine was proved by the long time it must have taken for the earth to change the texture of their skin. This contrasts with how dirty or calloused hands are typically read by *q'aras* as a negative marker of indigeneity, as captured by Rudi Colloredo-Mansfield (1998) in a moment of money changing hands in Ecuador. Miners have reworked the meaning of the matter in hand, investing earth-ingrained palms with the power to transform their owner into a true worker, or a man who had earned his right to lead others. I knew a teenage boy who accidentally hit his thumb with a hammer intended for the handpick he had jammed into the wall, and he practically glowed as he showed me his impossibly purple nail. He was only fifteen, and his hold on manhood was tenuous, but he treated his thumb like a ticket up. His individual bodily degradation was also the promise of joining an elevated social formation.

For women, however, calloused and earthy hands are not such an admirable trait and do not by any means guarantee access to leadership spaces. In the leadership transition that Rosa mentioned, the woman who won had given a speech about how the women's representative needed to be "well formed" (*bien formada*) to speak publicly in front of men. This comment was targeted primarily at her opponent who dressed *de polleras*, who had just given a powerful statement about how her knowledge of the mine would make her a strong leader even though she could not read or write. Among these women, *bien formada* signaled something quite different than it would have if it had been two men talking about their *formación*. It meant formal education rather than labor, and the kind of education in question was sociocultural rather than political.

In Bolivia formal education in Spanish-language schools has long been a key tool of colonial governance, since it regulated who could claim citizenship rights and who would be treated like a perpetual child; Spanish literacy has been a "marker of the racial, spatial, and political boundary between the white, male, urban *letrado* (literate citizen subject) and the dark-skinned, feminized rural indio (Indian peasant)" (Gustafson 2009b, 18).[12] Being *bien formada*, however, is about bodily presentation as well as literacy and language skills. The woman who won the FENCOMIN election was "well formed" by mestiza standards, which implies a specific beauty regimen and sartorial choices. In Bolivia, as elsewhere, clothing signals the wearer's social position but not in a way that allows an easy analysis of class and gender. The traditional clothing of Bolivian *cholas* is shockingly expensive, certainly much pricier than cheap Western clothes, most of which are imported from China.[13] Nevertheless, being "well turned out" references the assumption that a woman who can read, write, and speak in public—and who is therefore considered qualified for a leadership position—will be *de vestido* rather than *de polleras*. In other words, a well-formed woman will dress and look mestiza, *q'ara*, or otherwise non-Indigenous. Hands, clothing, and race are in this way bundled differently for women miners than for male miners, since being a working-class mestizo man requires hardened (rather than manicured) nails.

This does not mean that men's clothing choices are immune to raced and gendered readings. On another trip underground, I spent the day with two brothers, Jhonny and Sergio, who relentlessly teased their cousin, José, in a way that simultaneously racialized him, feminized him, and returned repeatedly to a discussion of his rubber boots. This started with Jhonny recounting a joke about José's first day of military service. The commander had asked him, "Do you know how to ride horses?" "Yes," said José. "Do you know how to use a gun?" "Yes," said José. "Do you know how to speak Spanish?" "No," said José. All three men had grown up speaking Quechua together in the ayllu Chullpa, but José had migrated to Llallagua later than his cousins and did not speak Spanish as fluidly. Fluency in Spanish is a marker of both *mestizaje* and masculinity, in part because men who did not learn Spanish at home usually learn it during their twelve months of mandatory military service—during which they are also introduced to state nationalism "in a highly masculine, hierarchical, and racialized context" (Canessa 2012, 220).[14] This is putting it lightly: in the military, punishments are meted out for speaking Indigenous languages, and recruits are regularly humiliated by being likened to women or queer men. This seemed to be the place from which Jhonny was speaking.

Continuing with his torment, Jhonny turned to me and asked if it was possible for a man to become a woman in Canada. "Well, yes—" I began, about to launch into a discussion of gender affirmation surgery, but Jhonny interrupted me. He turned triumphantly to José: "Look, you can go to Canada to have your surgery done!" Annoyed that I had been used to set up this kind of joke, I asked Jhonny why they gave José such a hard time. Without missing a beat, Jhonny declared that it was because José's boots were too small. Jhonny held José's boots up to mine—which were larger—and declared that José could be a petite woman in Canada. This remained a recurring joke for the rest of the day. José was marked not only as more Indigenous because of his linguistic skills but also as more feminine, a quality that was materialized in his small boots. Although he might be dressed as a miner, they kept insinuating, the boots showed that he was actually someone else, working in drag.

"In the Andes," Mary Weismantel writes, "race is indeed corporeal, but the definition of the physical self is extended beyond flesh and teeth to include the clothing and objects that extend, shield, and adorn the body" (2001, 184). Drawing on Judith Butler, she describes this as a performance of race, or an understanding of race that is much less dependent on biology than on daily practices. I would add that it is also a sensitively materialist approach to subject formation, in that it does not privilege living (biological) matter over apparently nonliving appendages. Clothes and calluses, both technically nonliving protective layers (calluses, after all, are accumulations of dead skin cells), contribute to the formation of miners in ways crisscrossed by racialized and gendered differentiation. These matters and their meanings are formed in intra-action—calluses, for instance, form in the rub between skin and rock, but they have different meanings on hands with differently gendered histories. Their weightiness, however, is not easily sloughed off.

ACIDIC WATER

Back in the *pauwiche* with Samuel and Rosa, we were wrapping up the *ch'alla*: all of us had large balls of coca leaves in our mouths, and the miners were changing into their muddy and slightly damp work clothes, which they keep in their *pauwiche*. They strapped on kneepads made of car tires and rubbed motor oil on their hands as a minimal protection against the moisture-sucking dust. We exited the *pauwiche* and climbed down eight ladders strung end to end, each separated only by a small platform. The wood of the ladders was soaking wet and slimy under the hand, and a couple of the rungs were missing—Rosa reminded me to always hold the sides of the ladder because the rungs were untrustworthy. As we neared the bottom, a terrible roaring noise greeted us. We were at level 685, only 35 meters (one level) below the *pauwiche*, but we had just descended below the natural water table. Huge electric pumps were running full-time to remove enough water to make mining feasible. The sounds of rushing water and the electric pumps were amplified in the enclosed space until they seemed deafening. Samuel led me to the pumps and insisted I snap a picture, but my camera struggled to focus in the dark and the photo wound up depicting tubes disappearing into a dark void.

As an economic activity, mining requires a lot of water (Budds and Hinojosa 2012; Kemp et al. 2010). Not only do mines tend to monopolize streams and aquifers that could otherwise be used for agriculture or human consumption, but the water they return to the surrounding environment is often so contaminated that it harms existing vegetation. Not far from Llallagua, geographer Tom Perreault (2013) has written about the downstream impacts of the state-owned Huanuni mine, where sludge-like water has filled the river valley with infertile sediment and trash. When he invited me to visit his field site in 2013, he pointed out that there were so many plastic bags deposited throughout the watershed that the fields seemed to sparkle from a distance. In contexts with stricter environmental regulations, mines usually discharge sediment-rich acidic water into tailings ponds lined with an impermeable substance. Tailings ponds, however, are far from failsafe: barriers can break or overflow, and when they do, they unleash toxic slurry across the landscape. This happened in Canada at the Mount Polley gold and copper mine in 2013 (Quastel 2017), and it happened in Brazil at the Córrego do Feijão iron ore mine in 2019 (Freitas and Silva 2019). Even in the absence of catastrophic breaches, mine waste left in ponds or heaps can slowly leach acidic water and heavy metals into the surrounding environment (Johnson and Hallberg 2005).

Across academic disciplines, it is widely agreed that mining has negative effects on regional freshwater resources, usually creating a combination of scarcity and contamination. Less commonly discussed, however, is the relationship between water and miners, or more precisely how water's intra-actions with rock and flesh leave permanent marks on the latter. As with other material intra-actions, these marks are not uniform across miners. Subterranean water, like the rock, is historically formed, materially specific, and deeply meaningful. These characteristics crystallize at the specific moments, and in the specific locations, that the water appears in the labor process. It matters who is doing the work at these times and in these places.

When we reached his *paraje*, Samuel began to shovel the *llampu*—fragmented rocks with flecks of embedded tin—into sacks, which he carried to Rosa, who began the process of selecting the ones with enough ore to justify hauling them aboveground. The few women who work un-

FIGURE 4.1. Rosa washing rocks underground. Llallagua, February 10, 2017. Photo by author.

derground are often assigned this task, which is deemed less physically strenuous than drilling, planting dynamite, or heaving sacks of *llampu*. It is also the subterranean equivalent to what *palliris* do on the surface, which involves sorting through and organizing rocks based on their ore content. The job of sorting remains feminized whether it is practiced by *palliris* or *mineras* and is a way of maintaining gendered hierarchies despite the transgression of women miners into subterranean spaces.

Rosa sat on a ledge with a giant plastic tub and a large hose of gushing water—it was the water, she explained, that had been pumped from down below up to level 650, from where it now gushed back downward with the help of gravity. She had a makeshift sieve, a plastic jug that had been cut open lengthwise and perforated with dozens of small holes, into which she poured a small amount of *llampu*. She washed the rocks clean with water from the hose, and what had looked like a mass of dirt became a clearly defined collection of stones, some sliced with dark streaks or peppered with crystals of tin. "Tin always reveals itself when it is wet," she informed me. Her hands moved quickly, plucking the valuable stones from the rocky chaff and dropping them into a burlap sack (see figure 4.1).

She paused to show me her hands, which were puckered and peeling, and chuckled as she asserted that you can always recognize miners from their hands. But *her* hands were specifically those of a woman miner, and even more specifically a woman miner who worked in a poorer *cuadrilla*. When rocks with high sulfide content are exposed to water and oxygen, they release hydrogen ions that substantially lower the pH of water (Lottermoser 2010). In the tin mine, the ore with the most tin—cassiterite—is a tin oxide that contains relatively little sulfide. But there are also many veins of tin sulfide in the Juan del Valle Mountain, and given the current exhaustion of the mine, more and more miners are exploiting these sulfurous deposits. The more they do this, the more acidic the water becomes. And since women are more likely to spend their days with their hands in the water, they are the ones who bear the brunt of this geological exhaustion. Gendering thus takes place at the intersection of flesh, water, and hydrogen ions; Rosa's body is formed as a woman worker specifically in relation to the slow degradation of the subterranean mine and the acidification of its water.

Rosa is a relative rarity in the Juan del Valle Mountain, since most women miners work on the surface, but water marks them there too. While all women who work on the surface are casually referred to as *palliris*, among themselves they are divided between "true" *palliris*—women who traverse slag heaps looking for rocks that they might break open for tin—and women who operate *budles*, or buddle pits. A *budle* is a circular basin carved into the earth, with a bundle of branches suspended vertically above its center. A slurry of water and ore sands runs along a narrow trough, down the bundle of branches, and into the center of the pit. Tin compounds collect in the center of the basin, while the lighter waste rock is washed by the water toward the basin's edges. Buddle pits are one of the oldest forms of "slime concentrators," having been developed in the Cornish tin mines at least as early as the thirteenth century (Crowfoot 1914, p. 1932). Indeed, German miner Georgius Agricola devotes considerable space to describing the buddle in his book *De re metallica* ([1556] 1950), thought to be the first modern treatise on mining (see figures 4.2 and 4.3).

In Llallagua most *budle* operators work at the base of the mountain, downstream from Llallagua. I got to know a *budle* operator named Adriana at a leadership workshop in town, and she invited me to come see how she worked. She built her *budle* with scraps of metal and branches that she harvested from the fields of abandoned machinery surround-

FIGURE 4.2. Buddles as featured in *De Re Metallica* (Agricola [1556] 1950, 302).

FIGURE 4.3. Operational buddle. Llallagua, July 21, 2014. Photo by author.

ing the Sink and Float Plant, after having learned the technique from her husband and friends. Using the same acidic water that pours out of the Juan del Valle Mountain and downstream toward her *budle*, she processes low-grade *mamo* ("sucked" ore, or sand that has already been processed once or twice) that she buys from other miners. She operates her *budle* leaning semiprostrate on her side all day, scraping a flat piece of metal back and forth across a screen through which the ore slurry flows. Small rocks collect in the screen while the rest sluices toward the basin. She adds xanthate, a chemical that releases tin from sulfur, and motor oil, which loosens up the waste rock. She explained that the water that exits the mine becomes more acidic in the sun, which also helps the rocks separate from one another. The acid becomes so strong, however, that by the time the water gets to the *budle*, "te raja" (it cuts you). Even more than those of her underground *compañeras* like Rosa, Adriana's hands are raw and peeling. Her body, like Rosa's, is being formed in relation to water and the degradation of the rocks. The more underground miners replace cassiterite with tin sulfides, the more xanthate and acidic water Adriana must use, and the longer she has to labor over her *budle* to salvage the same amount of mineral.

Subterranean water has its own history: it moves through the water cycle, it filters through the soil, and it can remain locked in aquifers for hundreds of years before humans crack its mineralized lid. Once freed, it reacts chemically with the many surface areas exposed by mining activities. In those moments in which miners make use of acidic water, it marks their bodies as much as it changes their rocks. It marks difference, degrading bodies even as it is forming metal.

DRILLS, MAKERS OF MEN

"You don't have any pictures of drills," Walter said, leafing through a stack of glossy photos I had printed for a poster we were making. A cooperative miner in his forties, Walter was serving a two-year term as the coordinator of health care for FERECOMINORPO, the regional federation of mining cooperatives in Norte Potosí, and he and I had been enlisted to make posters about the health risks faced by miners for Llallagua's health fair. María, host of the daily mining radio show *K'epirina* and unofficial liaison between cooperative miners and the rest of the town, had left us with a box of colored markers, poster paper, and pinking shears, instructing us to make a series of posters that she could use to decorate the walls of the kiosk we would be sitting in the next day.

Watching Walter's evident discomfort with the construction paper and markers, I had volunteered to run to the local photo shop to print some of the pictures I had taken while accompanying miners underground. Walter had enthusiastically agreed, but the photos I returned with were not entirely to his liking. He did not appreciate the stark lights of headlamps shining across rows of grim faces, or the close-ups of rocks and hands. "There has to be a picture of someone drilling," he said, tossing my photographs aside. "Maybe we can print something from the internet?"

Chagrined, I woke up María's computer and searched "Bolivia minero perforando" (Bolivia miner drilling). "That one there!" Walter crowed, pointing at a picture of two men standing in a tunnel, one of them directing a drill into the wall while the other appeared to watch. The picture, which we found reproduced on dozens of online fora, was low quality and looked blurry when we printed it from María's computer. "Better than nothing," said Walter as he ran a glue stick along the back of the photograph before pasting it to our poster. I tried not to feel bitter that my carefully curated set of photographs apparently amounted to less than nothing. No drill, no miner.

I had not yet taken any pictures of miners drilling in part because I did not relish the prospect of spending long stretches of time underground with billowing silica dust, unrelenting jackhammer noises, and minimal safety precautions. But avoiding drills had not been particularly hard since not all *cuadrillas* use drills. There is no electricity in work areas, and the drills rely on compressed air sent from air compressors kept in *pauwiches*, which are electrified. Air compressors are prohibitively expensive, costing around US$3,000. Although some *cuadrillas* share air compressors, even the cost of the drill (several hundred dollars) can become a financial burden if the ore in one's work area does not prove sufficiently high grade. For instance, Samuel is the *cabecillo* (head) of his *cuadrilla* because he purchased a drill, with which he was able to begin exploiting a new vein. The vein now belongs to him because he mixed both his living labor and dead labor (the drill) with nature, but in a neo-Lockean twist, he can hardly exploit it because his dead labor was repossessed by the financier. The vein was insufficiently rich to pay off the debt he had incurred by purchasing the drill, and he has returned to the classics: hammers, picks, dynamite.

Walter's need for a picture of a drilling miner therefore reflected less a universal *use* of drills than a universal *desire* for drills. Drills seem to produce miners as surely as they produce metal as a commodity. To some extent, a miner is only recognized as such when they know how to drill. Male miners frequently asked me, when I mentioned my mining women friends, if they knew how to drill. If I responded in the affirmative, the men would nod their heads, suitably impressed. Those women who did not drill did not fully count as underground workers, even if they descended every day. The drills and the subsoil spaces defined one another and the workers who intra-acted with both.

The following day, Walter and the other FERECOMINORPO representatives outdid themselves with drills. Whereas María had suggested that we find a mannequin and dress it as if it were a miner about to go underground, Walter produced a real miner (persuaded to skip work with the promise of free lunch) with a real drill, real dynamite, and a real rock for a drilling demonstration. We also set up a small statue of El Tío—the devil figure of the Bolivian mines who alternately promises riches and threatens catastrophes. We arranged cassiterite rocks, coca leaves, and *mistura*—colorful confetti used for blessings and celebrations—around El Tío's base. Finally, almost as an afterthought, we pinned up the posters we had made on the walls of the kiosk.

It was a health fair, and most of the other tables were occupied by representatives from the local hospital and the medical schools of UNSXX. These stalls featured droning presentations given by university students

FIGURE 4.4. Miners demonstrate drilling at the health fair. El Tío is in the foreground, the jackhammer on the ground, and our posters in the background. Llallagua, October 27, 2016. Photo by author.

that demonstrated how various diseases were transmitted, recognized, and treated. María, who had arrived to coach the FERECOMINORPO miners before she broadcast their public statements on her radio show, urged them to similarly describe their experiences with silicosis and tuberculosis. No one, however, was eager to dwell on such topics.

Instead, they were excited to demonstrate their drilling prowess. The miner they had brought posed with the drill for a series of photos with a rotating cast of university students. This drill was not actually operational since no one had wanted to haul an air compressor down from the mountain, but the FERECOMINORPO leaders had also brought a jackhammer that could be plugged into an electrical cable. With this, they took turns drilling into the rock they had brought, creating holes the size they would normally make for dynamite and correcting each other's drilling form. A large crowd of students formed around them, as dynamite was infinitely more exciting than tuberculosis slides. The miners went on to demonstrate how they prepared dynamite, slicing it in two and sliding a fuse into each half. They were so eager to continue the show that they lit the fuse and then stamped it out right before it reached the dynamite. María grabbed my arm and pulled me out of the way, just in case (see figure 4.4).

The phallic symbolism of the drill does not require much conceptual detail to unpack. In her study of the map featured in Henry Rider Haggard's novel *King Solomon's Mines* (1885), which charted the path to mythical diamond mines in southern Africa, Anne McClintock (1995, 1–5) has already brilliantly articulated the connections between white male penetration of the subsoil and colonized women's bodies. Mining might be considered the colonial activity par excellence, an enterprise that clearly yokes the extraction of (nonrenewable, nonfertile) nature to the exploitation of racialized labor forces and the violent gendering of both people and land. Indeed, in much of the Global South, the reverberations of colonial history might be expressed at the tip of a drill bit.[15] In a part of the world where nature is regularly deified as Pachamama, the gendered and racialized symbolism of a rock's degradation is even more overt. As Weismantel writes, "In the Andes . . . violent sexuality is inextricably connected to race; it is as though whiteness brought with it a phallus" (2001, 171).

As underground miners—especially those who can wield drills—take on the white/masculine role of violator, their words and actions constitute rocks in sexually racialized ways. Cooperative miners often describe geological formations that have yet to be exploited—which will generate the richest ore—as *roca sana* (healthy rock). In contrast, rocks that have already been brought to the surface and processed are *llampu* (Quechua for "soft, spongy, weak") or *mamo* (Spanish for "sucked"). These rocks are not waste in that they are usually reprocessed by other (lower-status) miners in *budles* or concentration plants, but they are certainly less valuable than healthy rock. Drills and dynamite are only used in healthy rock, which speaks to the phallic symbolism of the drill, while *llampu* and *mamo* are further processed in artisanal systems often run by women. Drills transform healthy (virginal) rock into sucked (degraded) rock, while at the same time conferring masculine whiteness on those who wield them—those who have the power to degrade.

This power extends beyond the subterranean. As in other mining towns, domestic violence—intimate, bodily, and psychological degradation—is a serious issue in Llallagua. The multigenerational combination of exploitation at the hands of (white, foreign, wealthy) capitalists, violently masculinist work environments, and alcohol dependency (which often has its own history, as mining capitalists ensured a constant labor supply through debt and addiction) has created dangerous home environments for women and children in mining towns worldwide.[16] In

Llallagua domestic violence was a frequent topic of conversation, mostly among women, who treated it as a regrettable but inevitable part of life. For instance, I got to know a young dentistry student who frequently visited me to discuss the beatings she was receiving from her boyfriend, a teenage tin miner recently returned from military service (the same one who, a year before, had been flaunting his injured thumb). The traumas of military service have a lasting impact on family life in rural Bolivia, where wives often embody all their husbands were taught to hate (Canessa 2012). In this case, I had known the young woman's boyfriend for several years, and I struggled to connect the dimpled kid of my memory to the volatile youth she described. She refused to leave him primarily because she was from out of town and his family had become her only local support system. Even becoming a miner does not protect women from abuse, and some of the worst stories I heard were from women who had ascended to positions of relative power within their cooperatives. Just as miners perform masculinity by inscribing it on ore bodies underground, at home they perform it through its inscription on human bodies.

Male miners, in turn, worry about their own abuser when they work underground. Hilario, the older miner who drives the trolley cart in the morning, also took me on several subterranean walking/crawling tours in between his scheduled shifts. Of all the miners I met, he is the one who talked most about El Tío. Statues of El Tío are located throughout the mountain, and miners pay their respects with gifts of coca leaves, lit cigarettes, and spilled alcohol, typically while requesting that he keep them safe from cave-ins while guiding them to mineralized ore veins. El Tío was made famous among anthropologists by Michael Taussig (1980), who argued that the devil was a personification of the workers' experience of alienation from their labor in the moment of proletarianization. Although June Nash ([1979] 1993, xxxvii)—from whom Taussig borrowed most of his ethnographic evidence of the Bolivian mines—later criticized Taussig's reduction of a "rich variety of symbols involved in the complex of rituals in the mines" into a single narrative, his interpretation remains extremely influential. What neither Taussig nor Nash mentioned, however, and what Hilario repeatedly underscored is that El Tío has also been known to rape miners underground. This, Hilario insisted, is why miners must always work in pairs or groups: when a miner works alone, he is liable to fall unconscious and wake up with his pants around his ankles and blood on his legs. According to Hilario, El Tío takes the form of a tall blond man to perform this act. While this description could support Taussig's interpretation of El Tío as the personification of exploitation, the figure's racialized traits (blond, tall) suggest that the long history of colonialism matters as much as or more than the transition to capitalism.

I later asked other miners about Hilario's story, and about half of them corroborated it, while the other half dismissed it as hallucinations brought about from exposure to subterranean gases. Lethal concentrations of methane, hydrogen sulfide, carbon monoxide, and carbon dioxide accumulate in enclosed subterranean spaces, particularly when chimneys (vertical ventilation shafts) are blocked. Produced by sources as diverse as gasoline or diesel motors, decaying wood structures, and the oxidation of newly exposed rocks, these gases have nowhere to go. They are a frequent cause of death underground, where they overwhelm a miner's lungs before an escape can be made. But, Hilario insisted when I asked about the gas hypothesis, El Tío and gases are not mutually exclusive. He readily agreed that El Tío might take advantage of the gases, or even create them, to facilitate his work. In this case, the characteristics that Hilario ascribed to the gases—invisible threats that could cause bodily harm—were not so different from those he ascribed to El Tío. As with El Tío, the penetration of gases into the miner's nose, lungs, and bloodstream could not be fought with muscle or even skill. All were forces that could quickly degrade an unprepared miner.

This story seemed to mark the limit of any attempts to become more masculine, or more mestizo-like, through work in the underground. Indeed, it would be difficult to imagine a more apt illustration of the gendered colonial relations that went into shaping the mine. The racialization of Indigenous peoples throughout Latin America was also, concurrently, a process of feminization and infantilization (Nelson 1999; Wade 1997). While twentieth-century Bolivian tin miners attempted to assert their value as politically conscious and masculine mestizos, they remained structurally dominated within a colonial economy helmed by q'ara men in suits. A drill might help a miner move up an internal hierarchy, might form him as a "proper" miner, but at the end of the day, it cannot protect him from getting raped by a white man in the dark. No one is safe from El Tío and his subterranean gases.

Jhonny, Sergio, and José—the two brothers and their much-tormented cousin I introduced earlier—are drillers. They were enormously proud of their drill and air compressor, which they had just recently purchased, and they were eager to show me their new machinery. I accompanied them with some trepidation, since drilling is notoriously unpleasant, but I was also curious about the process that Walter had insisted was so important to a miner's formation.

When it was time to begin drilling, Jhonny and José stayed with the machine in the *paraje* while Sergio climbed back up to where the air compressor was kept, in the electrified *pauwiche*. From up above us, he would send fifteen pulses of air shooting down through plastic tubing to power the pneumatic drill. After we pulled on our protective face masks, I followed Jhonny and José into the tunnel. They were anxious that I get a good angle to take photos, which was challenging since there was only room for two people in any part in the shaft; I ended up downhill from them, about eye level with their feet (see figure 4.5). When we were all finally settled, Jhonny folded a spare burlap sack and tucked it into his waistband, creating a small cushion that he used for bracing his hips against the drill. José crouched down to use his hands to steady and direct the tip of the drill. And we waited.

When the first one-minute pulse of air came, it felt like it shook the whole mountain. Immediately the air was full of billowing dust, my eyes were burning, and I had to remind myself to keep breathing normally—Jhonny had warned me that panicking would make it worse. Tiny fragments of rock ricocheted off masks, helmets, and faces. Jhonny had to throw all his body weight against the drill to make sure it stayed correctly angled, and his whole figure vibrated along with the drill.

Relaxing momentarily at the end of the first pulse, Jhonny took off his helmet to mop his brow. Dust coated every hair of his eyebrows and a couple of stray strands on his cheeks, all suddenly standing out in relief. He glanced up, saw my own dust-covered face, and laughed. He said that I should be careful not to get "la herencia de la mina" (the inheritance of the mine)—silicosis of the lungs. He touched his own chest and said, "I guess we all have this, as soon as we start working." Just as miners go inside the mountain every day, the mountain also slowly goes inside of them, settling into their bronchial passageways just as it once settled at the bottom of a Triassic lake, the humble origins of the mountain range.

FIGURE 4.5. Jhonny and José setting up for the first pulse of air.
Llallagua, May 24, 2017. Photo by author.

I thought about the inheritance of the mine through the next fourteen one-minute-long pulses of air, which collectively felt like much longer. Silicosis is not the same as black lung, which specifically describes coal miners' lungs. Instead, silicosis is caused by crystalline silica, an extremely hard rock typically found in quartz. My common sense had told me that the disease was the simple result of lungs filling up with dust, but that is not exactly accurate. It is the body's immune response to the dust that causes the symptoms. As silica particles are deposited in the alveoli of the lungs, they are ingested by macrophages, a type of white blood cell that attacks foreign substances. This initiates an inflammatory response. Silica particles are encased with collagen, causing fibrotic scarring and nodular lesions that leave the patient short of breath and vulnerable to other pulmonary diseases, like tuberculosis (Pollard 2016). This is why some people develop silicosis faster than others—it depends on their individual immune systems—and why silicosis can continue to worsen even after a miner has retired, since the immune response operates on a different timeline.

Miners, however, really do treat silicosis like an inheritance, or more precisely like a passport that should guarantee them access to all the benefits of national citizenship (including, most important, a pension: the monetized inheritance). Miners regularly talk about leaving their lungs in the mountain, giving their lungs to the nation, or sacrificing their lungs for the good of their families and communities. Legally, miners with silicosis can begin receiving their pensions at an earlier age, so there is a distinct financial advantage to an official diagnosis. But dusty lungs are about much more than money, as an official diagnosis is also a symbol of self-sacrificial labor and a mark of mestizo masculinity that was honored among twentieth-century miners.

Again, however, not everyone has access to this diagnosis and its attendant meanings. Nearly a year prior to my trip underground with Jhonny and his family, I had spoken with a doctor in Llallagua's local hospital who mounted three X-rays of lungs with various stages of silicosis for me to examine while we chatted about miners' health (see figure 4.6). The lungs were arranged in order of severity. Scarring from silica dust showed up like a thickening mist, gradually turning the space inside the miners' rib cages opaque. The doctor explained to me that rates of silicosis are much higher now than they used to be when the mine was run by the state. The state company used hydraulic drills, and the continuously circulating water produced mud rather than dust. But the

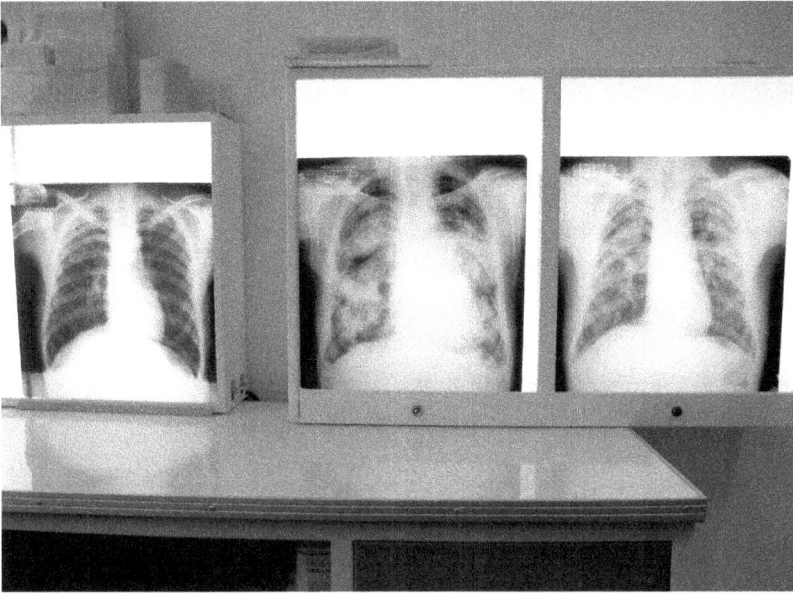

FIGURE 4.6. X-rays of cooperative miners' lungs on display at the local hospital. Llallagua, July 29, 2016. Photo by author.

small-scale cooperative miners cannot afford to maintain running water underground, so they use only pneumatic drills. "The literature says that silicosis should start after fifteen or twenty years working underground," the doctor told me. "But I regularly see miners with silicosis after having worked less than ten years."

I asked the doctor if he had any X-rays of women's lungs. He shook his head and informed me that women are not susceptible to silicosis. At first, he said that this is because women do not do any drilling, which is the task that generates the most dust. But when I informed him that I personally knew several women who had been drilling for more than a decade but could not convince his hospital staff to x-ray their lungs, he suggested that women have stronger bodies that fight against the impurities of silica. Apparently, scarred lungs are an attribute of "real miners" that women are not quite permitted to attain. Once again, the materiality of the underground interacts differently with different bodies, or at least is perceived to act differently in consequential ways.

Dust is similarly gendered on skin, even when it is visually imperceptible. When I was conducting an oral history with Demetria, one of the few women who—like Rosa—works underground, I was struck by her description of how her husband found out that she was a miner. When she started working, he had been hospitalized in a coma for more than eight months following a workplace accident coupled with tuberculosis. After attempting to pay his medical bills with sewing, cooking, and cleaning skills, and after she had already indebted herself to all her friends and family, Demetria was finally persuaded by her eight-year-old son to try her hand at mining in her husband's work area. Accompanied by three of her five children, Demetria entered the mine under the cover of night because she was not then a member of the cooperative to which her husband belonged—and in any case the mine was mostly off-limits to women. Typically, the eldest son will mine when his father cannot, but Demetria's eldest son was only eleven.

What happened to Demetria shows how the materialities of the underground create bodily differentiation even when the matter itself is invisible. Her husband was in a coma when she began working underground, and when he awoke, his doctors warned Demetria not to upset him, so she decided not to tell him right away. Because of his tuberculosis, he slept in a separate room, so it was easy for Demetria and her children to slip out without waking him. But despite all their efforts, he started to grow suspicious. When Demetria walked through the house, he declared, he could *smell* the mine. The mine has a very distinct smell—a composite of wet earth, dynamite, and the decomposing remains left behind by thousands of miners eating, drinking, chewing coca, and relieving themselves underground. Resistant to washing, this smell normally signals someone's respectable status as a miner, but on Demetria it was out of place. When her husband finally realized the truth, he was furious with her and disappointed with himself. As Demetria recounted:

> He told me, "I cannot accept this because I have always been a *machista* man, I have always believed that the mine was for men and not for women." . . . He began to cry, saying, "I regret all this happening to me and that you had to assume the role of mother and father. I never wanted you to enter the mine as a woman. I have to take on the responsibility as a man, I have to work, I have to support the house." But I had gotten used to working in the mine with my children, and I didn't want to be sidelined. . . . So little by little, we began working together, with our children, and we are still working all of us this way.

As she indicated, Demetria and her husband began to work together underground in one *paraje*, but it took him a while to feel comfortable with a wife who smelled, as did he and his *compañeros*, of the earth. Gendered bodies and gendered dust each have their own histories, and their intra-actions in the mine can have unsettling effects.

CORPOREAL ROCKS

My goal in this chapter has been to approach the topic of subject formation in a way that is attentive to how the practices and places of labor, as well as the material-meaningful qualities of nature, contribute to the production of social difference. Conceptually, I hoped to introduce an enlivened and meaningful sense of matter into historical materialist approaches to political formation, while also (re)introducing labor as an important site of analysis in approaches to "new" and especially geological materialisms. At stake in this project is the ability to think about how a host of nonhuman "things," each with their own histories and sets of meanings, can contribute not just to the formation of social worlds writ large but specifically to racialized and gendered differences.

Within Llallagua's mining cooperatives, labor involves both sorting tin from sedimentary waste rock and sorting out internal social hierarchies, a process that is always inextricable from divisions of race and gender. Within and across *cuadrillas*, above- and belowground, social divisions are cemented in relation to the geological properties of tin (cassiterite and tin sulfide) and the meanings that have been attached to its extraction—their material histories. Labors in the subsoil, particularly the act of drilling, are coded as masculine, mestizo activities that are only done "correctly" by those who mimic the practices of ex-unionized miners. Samuel and Jhonny, although they both have rural farming backgrounds, went to great lengths to frame themselves as miners in contrast to Indigenous campesinos who also happened to be mining. For Samuel, this distinction was marked by "dirty" rocks, or waste rocks discarded haphazardly around the work site, whereas for Jhonny the distinction was marked by rubber boots. Similarly, although some women can "transform" themselves enough to mine underground, their gendered transgressions are limited by the labors they are assigned and the dangers to which they are exposed (acidic water rather than dust from drills and dynamite). Their bodies are shaped correspondingly: for instance, Rosa's clean, peeling hands look dramatically different from Samuel's callused, dirt-ingrained hands. These bodily distinctions, acquired through specific labors with specific materials, crystallize specific social hierarchies. These ordering structures are constantly in flux, but what persists is the practice of ordering people and rocks in laboral relationship to one another.

Povinelli (2016) argues that the governance of the distinction between life and nonlife is a key feature of colonial late liberalism, a distinction that subtends the (more commonly interrogated) distinction between humans and animals. While the figure of the animal often appears in racial hierarchies authored in the West between biblical and eugenicist times, nonlife is already absented from these hierarchies. Apparently inanimate objects such as rocks typically only appear in racializing discourses with the "attribution of an *inability* of various colonized peoples to differentiate the kinds of things that have agency, subjectivity, and intentionality of the sort that emerges with life" from those that do not (Povinelli 2016, 5). In this chapter I aimed to show how the conceptual split between life and nonlife has a tenuous hold on reality. Material substances typically imagined as nonliving contribute not only to the constitution of life but to *differentiated* forms of life.

How does the degradation of ore bodies and fleshy bodies shape the political stances adopted by cooperative miners? In Marxist terms, how is consciousness shaped by nature's limits, by the exhaustion of non-renewable resources? No longer the subjects of revolutionary struggle, cooperative miners are still laboring, but their political projects are different from those of their unionized predecessors, different in ways that have been shaped not only by their lack of salaries but also by the material nature with which they wrestle. Rather than leaning into revolutionary struggle, they focus on satisfying immediate material needs. In national-level discourse, this trait is interpreted as political opportunism, but vacillation is also a symptom of geosocial precarity: it is a willingness to use any foothold on a slowly crumbling rock wall. And as the next two chapters show, these are not stable grounds for transformative political interventions.

5

INDUSTRIAL RUINS:
MATTERS OF TIME

Before one encounters any evidence of an urban population, Llallagua announces its presence through the *desmontes*, or slag heaps. From a distance, the *desmontes* look like a second mountain range running alongside the original—its afterimage, or its less verdant younger sibling (see figure 5.1). More accurately, the *desmontes* are connected to the mountains in the way that entrails are connected to a carcass. The guts of the Juan del Valle Mountain were not uniformly valuable, and the process of turning the mountain inside out yielded as much offal as meat. This connection is contained in the word itself: derived from the Spanish *desmontar*—to take apart—*desmonte* connotes the disassembly of the mountain (*monte*). The *des-monte*, the un-mountain.

The newly paved highway between Oruro and Llallagua snakes around a kilometer of *desmontes* before reaching the town. Small adobe and brick houses appear on the outskirts, their corrugated metal roofs held in place with rocks. Larger cement buildings near the bus station are decorated with graffiti expressing strong opinions about upcoming elec-

FIGURE 5.1. *Desmontes* (slag heaps) along the highway. Llallagua, July 8, 2016. Photo by author.

tions or unrequited loves. Directly across from the bus station, where it is immediately visible to any new arrivals, a mural features the faces of revolutionary miners who emerged from the region of Norte Potosí. The smallest figure—Maria Barzola, the only woman—looks upward at a huge portrait of Evo Morales. Above all of them is painted: "To those who with sweat and blood opened the path of change [*el camino del cambio*]" (see figure 5.2).

Just as this mural is more than a reminder of Norte Potosí's revolutionary history, Llallagua's *desmontes* are more than a ruined remainder of twentieth-century tin production. While the mural tells a tale of progressive revolutionary change, located in Norte Potosí's deep history and projected into a nationalist future, the *desmontes* both remind of past prosperity and contain a promise of future value. According to local leaders, these slag heaps still contain more than 18 million tons of low-grade tin ore that could be processed today, given an adequate infusion of new technology—though it is unclear what kind of technology, exactly, is required. In 2012 Evo's government granted half the slag heaps to the regional cooperative mining federation, FERECOMINORPO, an announcement that was received with much celebration in Llallagua and much dismay among political commentators, for whom it was further

FIGURE 5.2. Mural facing the bus station, with Evo Morales on far right. Llallagua, January 27, 2016. Photo by author.

evidence of Evo's duplicitous approach to environmental and labor politics. The gift was more symbolic than material since it was not accompanied by the technical or financial support necessary to exploit such a vast quantity of low-grade tin ore. Nevertheless, the long shadows of the slag heaps remain a tantalizing horizon of possibilities.

My central object of analysis in this chapter is what I call *industrial ruins*, a category that includes buildings, machinery, and waste rock left behind by the increasingly industrial mining practices of the twentieth century (approx. 1899–1985). I am interested in how people live with these ruins, a goal that I borrow from Ann Stoler's (2013) study of "imperial debris." Methodologically, this chapter relies on descriptions of walking tours I took with cooperative miners and other town residents, some of which I requested and some of which were pressed on me. By reflecting on what people wanted to show me and how they described what we saw, I explore how nostalgia and future imaginaries are unevenly materialized across the twin towns of Llallagua and Uncía. Infrastructure and machinery were never uniformly distributed, either above- or belowground, and this historical variation is further fragmented as it is reflected through residents' individual experiences; the stories show how residents relate to the ruins differently depending on their own raced and gendered positionalities. Put differently, the industrial debris is not only historical matter, as I have been exploring throughout this book, but also matter that is multiply historicized by the many people who live with and within it.

Yet there is similarity within the variation. Just as the apparently solid *desmontes* are formed of billions of discrete thumb-sized rocks, so too do all these disparate memories and imaginaries coalesce around a shared sense of arrested temporal progress. When it was first built, Llallagua-Uncía's state-of-the-art mining infrastructure was heralded by many as capable of transforming Llallagua from a remote hinterland into the nation's economic engine. Tin mining bankrolled the nation's economic development, while miners' unions promised political improvements through labor organizing and social struggle. Although mining management and miners might not have agreed about the precise content of the future—a divergence that shows up in the present—they all agreed that mining could unlock a future that was better than the past. This sense of time is complicated in the present, since the futures anticipated in the twentieth century failed to materialize as expected. Progress, such as it was, was detained by the crisis in the tin market and neoliberal restructuring in the 1980s. In the granular details of contemporary residents' interactions with the ruins, a commitment to industrial progress is paradoxically preserved long after the industry itself has collapsed.

Time also comes into this argument in a more specific way. One of the constitutive components of industrial ruins is the machines: worn out, broken, scattered into pieces, but still definitively there. Not only do these machines remind residents of the promise of progress, as do all the industrial ruins, but when the machines were operational, they *produced* the sense of time within which it was even possible to imagine progression on a linear axis. As Walter Benjamin ([1968] 2020) argued, the production of a modern sense of time underlies and is essential to the production of a modern sense of progress. Industrial machines, rather than following the cyclical patterns of sunrise and sunset, rainy season and dry season, or life and death, all created a linear, segmented, and potentially valuable sense of time. By extending the energies and capacities of miners, machines reduced the labor time required to produce a given value of tin, all while fundamentally transforming workers' specialized knowledges, bodily movements, and daily schedules. In this way, the machines of Llallagua helped produce a capitalist temporality, in which life was regulated by the never-ending process of creating value.

These same processes also created the material objects that are most visible on the landscape today, including not only the discarded rocks but also weathered buildings and broken equipment. Llallagua-Uncía might be considered a place abandoned or discarded by capital, but this is not what residents see. The questions of *who* sorts between waste and value, and what logics they use to make that decision, were not asked by Karl Marx. Yet these questions crucially inform how the ruins are interpreted—not as remainders but as possibilities. My core argument, therefore, is not only that industrial ruins are reminders of the promise of temporal progress but also that they disrupt historical narratives in which the living and nonliving contents of Llallagua-Uncía are figured as nothing more than the wasted remains of a more glorious era. By holding firm to the sense that there is still more value to be generated, residents of Llallagua-Uncía resist the notion that they might have been left behind by the nation's march through historical time. This conviction also motivates cooperative miners to keeping digging deeper, down through the geological strata where their future wealth might still be stored.

MACHINES, MINING, AND NOSTALGIA

The process of ruination is a social as much as a material process. As Ann Stoler (2013) argues, the material rot of colonial infrastructure leaves psychological scars in its wake, transforming everyday life in ways that are not always immediately obvious. Ruins have local meanings that align imprecisely with national or global narratives, showcasing lost futures as much as past failures. Imperial ruins "pulsate" in time with local stories, which means that they might conjure sensations of horror, guilt, pride, relief, and so on, depending on how the past is locally compared to the present (Gordillo 2014).

While it would not be much of a stretch to say that Llallagua's ruins are imperial debris, given the extent to which US and European empires stimulated the tin market and manipulated Bolivian mining policies, it also matters that these ruins are industrial machines rather than, for instance, colonial churches. When they were built, the machines and buildings I describe in this chapter sedimented a specifically twentieth-century vision of a technologically advanced future. Sheila Jasanoff (2015) describes this vision as a "sociotechnical imaginary," noting that such imaginaries are typically underwritten by narratives of social progress. This means that the ruins evoke, for most people in Llallagua-Uncía, a positive collection of emotions. The ruins are objects of nostalgia, but what is yearned for is not the past so much as the imagined future that might have been and (perhaps) could still be.

In addition to these ruins consisting mostly of industrial machines, it also matters that they are specifically geared toward mining. Since at least the sixteenth century, the relationship among machines, mining, and narratives of progress has been highly ambivalent, since mining is necessary to yet distinct from industry as it is usually conceived (in urban centers, on factory floors, etc.). On the one hand, metal is *the* key ingredient of most "modern" machines, which means that technological advances in other sectors always result in intensified mining. Railroads, steamboats, and tractors all relied on excavation of metal ore as well as carbon-based fuel. Miners, moreover, whether in England, India, or Bolivia, used both their power as workers and their power as symbols of progress to agitate for improved social conditions. In this way, mining, machines, and social progress were linked together in the minds of both mainstream and Marxist-socialist development advocates. On the other hand, mining is a truly ancient practice, and in Latin America (as in much of the world), mining is inescapably associated with the socially and environmentally devastating legacies of colonialism. Moreover, mining as a sector was surprisingly slow to mechanize. Prior to the twentieth century, most mining worldwide was done manually with chisels, picks, and pans, even though it was so integral to mechanized industrialization around the world. Dynamite, invented in the 1860s, was not widespread for another several decades, and pneumatic drills appeared around the same time (Mumford [1934] 2010).

For mining, it is not just the technological symbolism of machines that is important but also the colonial symbolism of ripping up and transforming nature. In the early colonial era, European philosophers and theologians evaluated the "sophistication" of other communities in reference to their apparent ability to transform or otherwise dominate their natural material surroundings; the disciplining of external nature was conceptualized as an integral component of mastering one's internal (human) nature. This was the attitude with which Spanish conquistadores opened metallic veins across the Americas, most notably in the silver mines of Guanajuato (Mexico) and Potosí (Bolivia), while mandating religious education through which Indigenous Americans were to learn control over their internal "animal" natures. In the eighteenth and nineteenth centuries, as God was displaced by science, technology was increasingly used as a proxy for evaluating civilizational "progress." In other words, machines became "a measure of men [*sic*]" for Europeans busy justifying colonial expansion (Adas [1989] 2015). According to this informal metric, the bigger and more complicated the technology, the more advanced the community. For instance, British philosopher and eugenicist Herbert Spencer argued that human consciousness was reflected in technological transformations of nature, while British colonial officer H. H. Johnston argued that "there is no civilizer like the railroad" (quoted in Adas [1989] 2015, 229). Framing themselves as technological ambassadors of global progress, European and increasingly American forces expanded their colonial claims throughout Africa, Asia, and the Americas.

Faith in technology, however, was not limited to capitalist or western European contexts. In the twentieth century, as former colonies declared their independence, "technology transfers" became a key weapon used by both Soviet and American forces in Cold War efforts to win allies among the nonaligned Third World. Latin American countries, although most had been officially decolonized for a century and a half, became particularly fierce Cold War battlegrounds. Following Bolivia's 1952 National Revolution, the USSR offered to provide the country with a new tin smelter and US$150 million in economic and technical support (Field 2014, 12). Up to that point, Bolivia had had to ship all its tin to the United Kingdom and the United States for smelting, which meant that the country was constantly losing money by selling unrefined ore. To counter the Soviet offer, the US Kennedy administration partnered with the West German government and the Inter-American Development Bank to collect funds for the Triangular Plan, as discussed in chapter 2. Later still, Czechoslovakia funded the construction of a smaller antimony smelter, which was finished in 1975. While the Cold War powers each tried to lure the Bolivian state with machinery, the state, in turn, passed along used machines to the cooperative miners, often with a heftier price tag than they deserved—a dynamic that was described to me by a La Paz–based historian of gold-mining cooperatives (interview, La Paz, March 3, 2017). Regardless of the benefactor, gifts of metal promised to win the future.

Soviet commitments to technological change can to some extent be attributed to Marx, who hinted toward what was later described as "technological determinism," or the idea that a society's technological capacities determine its social structure. Most famously, in *The Pov-*

erty of Philosophy, Marx wrote, "Social relations are closely bound up with productive forces. In acquiring new productive forces, men change their mode of production; and in changing their mode of production, in changing the way of earning their living, they change all their social relations. The hand-mill gives you society with the feudal lord; the steam-mill society with the industrial capitalist" (Marx [1847] 2008). According to this passage, the productive forces of society define the contours of human social relations, but Marx was somewhat ambiguous about what exactly he meant by "productive forces." Although the phrase is often interpreted simply as technology (Heilbroner 1967), there is good reason to think that Marx meant a more complex arrangement of people, resources, and machines. For one, Marx did not think of machines as inert objects but rather as "dead labor," since they congealed past human labor and parsimoniously dripped it into manufactured commodities. Moreover, he understood the development and expansion of technology within the labor process not as the cause of social change so much as an expression of antagonisms already inherent within capitalist relations of production. Machines, as dead labor, continuously displaced living labor by dramatically reducing a capitalist's need for workers, and they reduced the worker from an artisan to an appendage.[1] In this context, workers' sense of alienation from the products and processes of production should develop "into a complete and total antagonism with the advent of machinery" (Marx [1867] 1990, 558)—an antagonism that would ultimately resolve into transformative social change. This is still a teleological process, but it is less determined by technology so much as internally bound up with technology.

Multiple narratives of progress thus condense in the rusted hulls of metallic giants: progress through the transformation of nature, progress through technological development, and progress through collective social action. Viewed in this way, the windswept ex-industrial landscape is a graveyard of hinges that once articulated the towns of Llallagua and Uncía with national- and world-historical temporalities. Yet apparent ruins also murmur promises of future value, while obsolete machinery reignites technological optimism. Materialized above- and belowground, nostalgia continues to impinge on imaginaries of the future and inflect the politics of everyday life.

Although the interior space of the Juan del Valle Mountain is one vast network of interconnected tunnels and mine shafts, the mountain's surface is divided into two municipalities: Llallagua, on one side, and Uncía, on the other. Each of these towns, in turn, is composed of several sub-municipalities, two of which are more powerful than the others: Catavi and Siglo XX, both located within Llallagua. All these four urban areas—Uncía, Llallagua proper, Catavi, and Siglo XX—were originally separate mining camps established by different prospectors in the late 1800s, but by the 1920s they were unified under Patiño Mines. Each region was the stage for a different piece of the company's operations, and each retains its unique architectural and social character today.

UNCÍA

During the early months of fieldwork, I spent most of my time in Llallagua. Uncía is smaller and slower, and far fewer cooperative miners enter from the Uncía side of the mountain. But when I told this to Toby, a former cooperative miner in his forties whom I met outside Don Alfonso's office at FERECOMINORPO, he furrowed his brows and insisted that I had it all wrong. He offered to take me on a tour of Uncía, where he had grown up and where he thought the region's "real history" might be found.

At first blush, Uncía looks very similar to Llallagua. The buildings have the same plaster walls, and the streets are similarly steep and paved with cobblestones. The main difference I noted was the larger number of women dressed *de polleras* (with the layered skirts that signal urban indigeneity), which aligned with what middle-class shop owners in Llallagua had told me (often disparagingly) about Uncía—that it was the town of choice for families migrating from the ayllus to work or to send their children to school. Toby, however, drew my attention away from the bustling market and upward toward the building rooftops. "Look at that wood!" he said, pointing at a row of wooden balconies jutting out into the street from above, and then to heavy doors and window frames, all of which were made of the same deep mahogany-colored wood. "No one builds from wood on the altiplano. Where would they get the lumber? It was probably imported all the way from Spain. That's how you know that Uncía was home to all the millionaires." He ran his hand over a time-smoothed window ledge admiringly and went on to explain what each of the wood-trimmed buildings had been turned into: the city hall, a hotel, a theater. "It makes Uncía prettier than Llallagua, don't you think?" He did not dwell on the things that caught my eye—the places where the wood had cracked or broken away, or where it incongruously abutted simple cement structures—and he seemed to find the idea of importing European building materials to the middle of a landlocked South American country impressive rather than excessive.

Indeed, it occurred to me that Toby's current job was like the modern-day equivalent of bringing wood to the altiplano. Since retiring from the mines, he had started a business laying down artificial turfgrass for soccer fields in towns around the country. By far his biggest customer was the MAS (Movement toward Socialism) government, since building soccer fields was an easy way to demonstrate a commitment to rural development. Perfect green squares of plastic grass can now be found in nearly every altiplano town, lurid against the backdrop of the muted grays, yellows, and reds that dominate the mountainous landscapes. Toby's business was booming. Whenever I met him for lunch, he spent most of the time scrolling on his phone through pictures of the property he had recently purchased, a sesame seed farm in the Santa Cruz lowlands with a ranch-style house and a kidney-shaped belowground pool. His admiration for the industrial modernity represented by Uncía's wooden embellishments seemed consistent with his turfgrass present and his agribusiness future.

Once satisfied that I had appreciated Uncía's downtown architecture, Toby led me upward, where we passed the old train station. Although it is now a private residence, railroad tracks still run alongside it, and ESTACIÓN UNCÍA is still painted above the door. This station dates to the 1910s, when Simón I. Patiño, impatient with the Bolivian government's efforts to build a national rail system, requested permission to build this stretch of railroad himself. When completed in 1925, it connected Uncía to the nearby town of Machacamarca, from where trains continued to Oruro and eventually to the Chilean coast (Oporto Ordóñez 2017, 96). Prior to its construction, Llallagua and Uncía had been connected to Oruro only via footpaths, and tin was transported on the backs of humans and llamas. Today this station—like most others around the country—has been shut down. The rail system was privatized starting in 1995, whereupon it proved unprofitable in the absence of a robust mining industry.

From the station, Toby and I continued up a very steep hill surrounded on each side by small houses that Toby said had been built in the past decade by "people from the countryside," which is code for Indigenous people from the surrounding ayllus. Toby, whose family had lived in Uncía for generations and whose father and grandfather had both been miners, insisted that the newcomers did not respect the town's history. It was this history, rather than the present, that he wanted me to see.

With this in mind, Toby led me straight up to the highest part of the town, where Patiño had built his personal residence. In truth, it was only *one* of his personal residences since he also built estate-like houses in Oruro and Cochabamba. Like those other two, the Uncía house had been converted into a museum dedicated to the mining industry. It was a striking building, perched on the edge of the cliff with one tower rising like a church spire. It had recently acquired a coat of yellow paint with white trim, which contrasted starkly with the dilapidated buildings surrounding it.[2] These had been part of the company's *gerencia* (management)— business offices, bakeries, butcher shops, and so on—and all of them had been left to continue their multidecade decay undisturbed. According to Toby, Patiño had designed the house to maximize his surveillance of mine shafts. From the round window in the spire, he could easily monitor who—and what—was going into and coming out of the mountain. At the top of the tower was also a giant clock, which Toby explained used to ring out three times daily, to mark the transitions between the workers' three shifts: at 7 a.m., 3 p.m., and 11 p.m. The mine never slept, and an equal number of workers showed up for each eight-hour shift. I stared at the yellow spire for a long while after that, marveling that the one building could have disciplined the workers in so many ways, ensuring that work happened when and how Patiño desired.

There was no one in the museum to charge us the five-boliviano entry fee, but the door was open, so we left our payment next to our scribbled names in the guestbook. Most of the exhibits were photos scrounged from archival footage: pictures of miners going in to work, of foreign engineers standing next to newly opened mine shafts, and of Patiño standing outside his house with his wife and daughters. The whole family regarded the cameras seriously, the women dressed in starched white dresses that must have been difficult to preserve in the dusty mining environment. Other than the photos, the only objects on display were pieces of machinery. The machines were unlabeled, which meant that to my eye they were inscrutable, but Toby recognized each piece. When I expressed interest, he insisted that he knew where we could find the "real" historical machinery, and it wasn't in the museum.

He dragged me out the door and down the hill to two giant processing plants, each of which looked thoroughly abandoned. The cathedral-style windows were either grimy or broken, and the edges of the building were overgrown with shrubs. The door was padlocked shut, but Toby boosted me up with both hands so that I might peer in one of the cracked windows. Visible between the jagged edges of glass were all the machines that Patiño had used to transform rock into mineral, arranged on the floor exactly as they had been when operational. They had clearly been assembled inside the building, or maybe the building had been erected around them—in either case, there would have been no way to move them out when the company shut down. They made the pieces on display in the museum look puny in comparison, which was clearly what Toby wanted me to observe: "It's impressive, right?" he called up to me. And: "Don't forget to take photos!" I obliged.

Toby saw in Uncía what he had been taught to see by his successful mining family: wealth, grandeur, innovation. Against a narrative of dispossession, contamination, and waste, he insisted on remembering the town's halcyon past. Promises of progress past, whispered in and through the material infrastructures of his town, continued to inform his actions in the present.

LLALLAGUA, CATAVI, AND SIGLO XX

If Uncía was where the wealthiest segment of Patiño Mines' management lived, the mining "professionals"—engineers, geologists, doctors—all lived in Catavi, now a submunicipality of Llallagua. Bolivia did not have a robust higher education system for mining professionals in the early 1900s, so Patiño lured foreign expertise from other companies around the world. The first time I went to Catavi, it was in search of the archives, because all the files connected to Patiño Mines had recently been relocated from El Alto into one of Catavi's abandoned houses. Although the houses in Catavi lacked the wooden flourishes of Uncía, they were larger than any of the buildings in Siglo XX or Llallagua proper and were designed in a Spanish style, with open courtyards at their centers. A low fence ran around the perimeter of the building where the archives were housed, and the imposing front door was bolted shut. Inside, the trees in the overgrown courtyard still conveyed a sense of opulence, and the large windows let in enough sunlight to combat the frigid weather.

Lidia, an archivist who had grown up in Llallagua but worked in La Paz for many years, oversaw Patiño's documents. She had hired a small team of assistants, including a nursing student and a former cooperative miner, to help her with the process of separating still-legible papers from those that were too rotten or brittle or moth-eaten to preserve. Most of these papers had been collected from other houses in Catavi, where they had been abandoned—along with the houses themselves—after the mine shut down in 1985 and the professionals all returned to their home countries. Although Lidia and her team had made significant progress by the time I showed up, they still did most of their work with masks to protect themselves from the dust and mold that covered everything they handled. Piles of unsorted books and papers were stacked in every corner of the creaky building, and collections of maps littered the floor.

I spent several weeks with Lidia and the other archivists as I examined employee files from the 1910s through early 1950s—that is, workers based at Patiño Mines before it was nationalized. I reviewed more than six hundred files, but I could easily have continued for months; there were thousands to pore over. Each employee had a folder marked with their number, and inside each folder was a pink sheet of cardstock where essential details had been punched in with a typewriter: name, age, birthplace, year hired, job(s) held (mechanic, day laborer, peon, driller, etc.), department(s) (milling, concentration, workshop, interior mine, etc.), daily wage, and year of termination. Behind the pink cardstock were all the documents corresponding with the person's employment, including layoff notices, medical reports, and letters from widows requesting support after their husbands' deaths from silicosis. These letters were usually accompanied by carbon copies of typed replies explaining why the company could not be held liable and offering small monetary "gifts" rather than suitable compensation. Of all this information, the most arresting aspect were the workers' photos, which were glued on alongside thumbprints and signatures (the latter of which was often replaced by a curt *no sabe*—"does not know"). With one photo taken from the front and another from the side, and their employee numbers behind them, these photos look exactly like prisoners' mug shots. Unsmiling, the men in these pictures seemed Dickensian to me, with jackets layered over sweaters over collared shirts, scarves bound tightly around their necks, and floppy pageboy hats pulled over their ears. Some faces were shockingly young, barely teenage, whereas others were etched with deep lines.

The material presence of these folders and photos in Catavi is important for locals, since they are all that remains physically of the workers who were just regular employees rather than union leaders or martyrs. Several times while I was working in the archives, people showed up asking if any information could be recovered about their grandfathers or great-grandfathers. Armed with names and sometimes employee numbers, the archivists would do their best to locate the folder in question, though the sheer number of files and their relative disorganization made it a challenging task. If found, these records could sometimes be used in legal struggles for property or pensions; other times, their importance was more symbolic. Claims to local and national belonging, legal or otherwise, were caught up in these folders.

In addition to helping me out with the files, the archivists also treated me to informal walking tours of Catavi when we made our way home each day. They pointed out a massive hall that had once housed an exclusive club for engineers and managers and is now rented for weddings and other community events. The old workers' hospital had been acquired by the university to teach nursing students, and young students stream in and out at all hours of the day. A series of pools fed with natural hot springs, once the exclusive domain of managers and their families, is now open to the public and constantly overflowing with guests. All these buildings, most grown shabby from lack of maintenance, retain something of their stately past in the rounded shape of their windows, decorative tiles, and the oddly temperate-climate trees around their perimeters. Although repurposed, these historical details are clearly valued by locals, who went out of their way to tell me about the golf course that Patiño had built to host visiting businessmen, or the airplane that arrived weekly with fresh meat for the *pulpería* (company store). To this day, people who live in Catavi are the middle and upper class of Llallagua—which is not to say that they are wealthy, only that they are comparatively so.

On the other end of this wealth spectrum is Siglo XX (Twentieth Century). Named after the mine shaft that it was built around, Siglo XX was and continues to be a working-class enclave. Some of the miners' original adobe houses still stand, and these small, one- or two-room windowless buildings are today occupied by the poorest of cooperative miners. Most of the houses, however, dissolved back into the earth when their sheet metal roofs were removed and repurposed following the great worker exodus of the 1980s. Rising only a few feet off the ground, the outlines of former walls maintain the houses as ghostly presences. Leading me through this patchwork of former buildings, a miner once explained that these houses had been huge improvements for early twentieth-century miners. Before these adobe dwellings, miners had lived in caves dug into the sides of the mountain. They worked underground all day only to sleep in a different subterranean space at night.

As the workers' home, Siglo XX was also home to the workers' movement. The union building, erected after the revolution of 1952, is in the Siglo XX plaza. These days, it is mostly abandoned, save for a regular contingent of elderly miners who file in to demand the pensions they are due. A faculty member from the local university took me to see the union building's pockmarked exterior, which he explained was the result of bullets fired during the massacre of 1967, when soldiers shot a controversial number of miners (estimates range from twenty to a hundred) during the dead of night. The soldiers arrived during the revelries of San Juan, an annual celebration that marks the coldest day of the year, and they used the noise of firecrackers as cover for their attack. In 2017 I joined cooperative miners and hundreds of others commemorating the fiftieth anniversary of the massacre, which involved a parade that snaked down from the mine shafts above to the Siglo XX Plaza. Several officials had arrived from La Paz to participate, including Llallagua's very own José Pimentel, then president of COMIBOL and former minister of mining and metallurgy. These men all gave speeches from the building's balcony, and although no one listened to the content—everyone was hungover from San Juan celebrations the night before, and we were sweating in the midday sun—it was significant that the officials had shown up in Llallagua and were speaking from the old union building. After the last person had spoken, everyone raised their left fists in the air to honor the fallen martyrs (see figure 5.3).

Each of these urban areas—Uncía, Catavi, Siglo XX, and Llallagua proper—has a unique historical weight to its residents. The weight is material: the architectural differences mark social differences that were established spatially as much as economically in the first half of the twentieth century. In this sense, the buildings were a technology in their own right, in that they broke the mining community into four parts, each of which contributed to a different constitutive aspect of the company. Altogether, the company operated like a single machine, partially submerged underground and partially sprawling across the surface of

FIGURE 5.3. Fiftieth anniversary of the San Juan Massacre. Llallagua, June 27, 2017. Photo by author.

the mountain, each segment of the machine working in harmony with the others.

Yet this apparent equilibrium relied on the creation of infrastructural (and other forms of) inequality between the urban areas, as well as even more substantive inequality between the urban areas and their rural surroundings. These infrastructural divisions segregated people along class lines, and these lines—although blurred—are retained in the present. This segregation allows some people to walk through Uncía or Catavi and reminisce about past economic grandeur, while other people visit Siglo XX and recall past political unity, when the working class contributed to the upending of not only the company's but also the country's socioeconomic structures. The tin-mining sector promised both economic and political development, and these different promises are whispered in wooden balconies and bullet holes, in museums and disintegrating adobe houses. Whether they are carefully maintained or left to the elements, these buildings remain a kind of bedrock on which the more contemporary community—young university students, Indigenous families migrating from their ayllus, and hundreds of cooperative miners—build their lives.

On the edge of Llallagua, along the blurry line that divides the town from the looming *desmontes*, is a large structure that appears fully abandoned. Constructed of sheet metal pockmarked with rusted holes, it is nearly camouflaged against the gray-and-rust-colored *desmontes*, noticeable only because it is set apart from the town and because it is frankly enormous. It is the Sink and Float Plant, constructed by Patiño Mines in 1947 to process lower-grade tin ore (Espinoza Morales 2010, 100) (see figure 5.4). At the time that it was built, it was considered the company's crown jewel, and it was regularly featured in black-and-white booster reports. Attached to the plant was an *andarivel* (suspended cable cart system), which transported waste rocks farther afield and facilitated the growth of the mountainous *desmontes* that currently circle the town. Today the *andarivel*'s cables are all broken, the carts are scattered on the ground, and the posts stand like sentries with nothing to communicate (see figure 5.5). Yet the *andarivel* in particular remains an object of pride for Llallagueños,

FIGURE 5.4. Sink and Float Plant amid *desmontes*. Llallagua, March 1, 2016. Photo by author.

FIGURE 5.5. Abandoned *andarivel* with *desmontes* in background.
Llallagua, February 9, 2017. Photo by author.

several of whom informed me that it was prototype for the *teleférico*, a
gondola-based public transit system that opened in La Paz in 2013.

I took the Sink and Float Plant to be valuable only as scrap metal until
I noticed its minimal but constant military presence. Tin is still classified
as a strategic metal, which means that tin-mining machinery still needs
protecting. Then I noticed that part of the structure is still functioning.
Although much of the largest equipment has been left to decay, the basic
internal structure of the plant is operated by Siglo XX, the largest cooperative in the region.

Mateo, a member of Siglo XX in his late thirties, was my guide through
the Sink and Float Plant. He began the tour by explaining that the guiding principles behind any concentration plant is *triturar, pulverizar, y con-
centrar* (grind, pulverize, and concentrate), and each of the machines
operated by his cooperative fell into one of those categories. From a spin-
ning mill that can crush up to half a kilogram of ore in each rotation, the
rocks are moved to a series of shaking tables, which are essentially vi-
brating inclined surfaces across which ore slurry sluices continuously. As
the tables shake, heavy tin particles slide the farthest across the table,
creating a thick black line of mineral grains that are harvested as they
pour off the left-hand corner of the table. To the right of the line of tin

is a thicker line of orange pyrite that is collected and directed to another shaking table lower down for a second round of processing. Any grains that fall to the right of the orange streak are waste particles and are directed straight into the river. In the river these tiny grains wreak havoc on downstream agricultural communities, where sediment buildup changes the river's course and acidifies irrigation water.

These shaking tables were imported by Patiño in the early twentieth century, and they are now encrusted with hardened rock crystals. Much of the nonessential framework of the building has giant holes rusted into it, only some of which have been covered with planks of wood. Yet Mateo kept describing the whole operation as *bonito* and *hermoso* (pretty, beautiful) and expressing excitement about how much each machine could process. With the eye of an antique collector, he blew dust off their worn-out labels to see when and from where they had arrived. He was satisfied that they were mostly American, with a few German pieces, and noted with pride that this was some of the most advanced mining technology in the world in the first half of the twentieth century. He showed me the joints on the bottom of the tables, which were smelted into solid pieces, and explained that Bolivian companies were now manufacturing shaking tables with joints that were simply welded together and easily came apart after a few years of continuous vibration. He shook his head in disgust at the quality comparison.

From the shaking tables, the ore moves to flotation units, which were Patiño's pièce de résistance when he first constructed the plant. The flotation units were how he would separate tin from pyrite in a mine with dwindling levels of cassiterite. Mateo explained that a reactive chemical (xanthate) and an oil is added to each unit once it is filled with ore slurry. As the unit spins, the oil binds to the pyrite, creating air bubbles around it and forcing it to rise to the top, where it can be skimmed off like a shiny metallic bubble bath. The strong chemical smell burned the inside of my nose as I leaned over a large hole in the floor to peer inside the unit. "Bonito, no?" (Pretty, isn't it?), Mateo repeated.

Although most cooperative miners expressed admiration for this old metallurgical equipment, Mateo's love was more pronounced. He was also wealthier and had received more formal education than most cooperative miners, and these details showed up in both his appearance and his approach to mining. The first time I met him, while waiting to go underground at the Siglo XX mine shaft, he was wearing a turquoise kaffiyeh wrapped around his neck, and he smiled at me with a row of perfectly straight white teeth—a true oddity in the mines. Interviewing him later, I found that he is a practicing lawyer who decided to get into mining on the side, since "that's where the real money is, as long you do it right and invest in machinery and technology" (Llallagua, September 26, 2016). Mateo became a miner in the cooperative Siglo XX to get to know the process and the other miners, but his real interest was metallurgy. In fact, with his father's financial assistance, he had assembled the machinery necessary to build his own small, borderline-illicit processing plant on the river downstream from Catavi. He made a small killing buying and processing low-grade ore that he purchased from other miners in his cooperative. "It's all about investing capital," he told me earnestly. "That's what this government doesn't understand. They're stuck in socialism, an old ideology that comes from outside [of Bolivia]. You need capital to make things productive."

After touring the Sink and Float Plant, I also visited Mateo's personal metallurgical plant in October 2016. I met him at his parents' house in Catavi, and from there we took his motorcycle. We had to cross a field of *colas arenas* (sand tailings) to get to his plant. Unlike the *desmontes* in Llallagua, which are formed of comparatively large *gramsa* (rocks about two to three inches long), the sand tailings are made of just that—sand. It's a shocking landscape. All the buildings disappear, and the panorama is reduced to a riverbed framed by bright red mountains on one side and bright white mountains of sand on the other side. The altiplano always reminds me of an overexposed photo, like the land itself has been bleached by sun, but here it is far more extreme. Down the middle, the river runs beige, full of mining waste that flows out of the mine shafts Cancañiri and Siglo XX and through all the small mining operations around town before snaking through this part of Catavi.

His processing plant, when we saw it, looked like a miniature replica of the Sink and Float Plant. The principle behind processing plants, always built on the side of hills, is to move the mineral from the highest levels to the lowest levels, using gravity to drive purification. Mateo was eager to show me how everything worked. We started our tour at the lowest level of the plant, where there were two shaking tables set up. One of the tables was American, and the other was Russian, and Mateo told me that the American one was much better because it shakes in a slightly oval shape, whereas the Russian one just moves forward and backward continuously. He proudly dusted off the label on the American one to show me that it was called "The Deister" and had been made in Indiana in 1906. Patiño had brought these shaking tables to Bolivia for tin concentration, but Mateo suspected that they had been designed to process gold originally. I later found the Deister website and saw that Mateo had been right: Emil Deister was an immigrant from Germany to the United States who designed the shaking table for the Alaska Gold Rush (Deister Concentrator Company 1906). Histories of extraction across Bolivia and Alaska were entangled not just by goals of profit and progress but also by the physical technologies used to achieve these goals.

Mateo went on walking up the levels of his plant, essentially moving backward through the concentration process. On the middle level were more shaking tables, and further up was the mill, a huge metal barrel filled with loose steel balls. When it rotated, the clanging balls slowly transformed rocks into sand. Above the mill was a *chancadora* (crusher) that accepts larger rocks and spits them out as smaller ones to go into the mill. Next to the *chancadora* was the pile of large rocks that had just been delivered from the mine. Mateo leaned over and started picking through the rocks to show me the different kinds of tin. Most contained *punti-tas* (tips) of tin sprinkled like pepper across predominantly white or gray rocks. He explained that for miners without access to industrial plants like Mateo's, these rocks were *mamo* (waste) because they did not have enough tin to justify processing in more artisanal systems.

We circled back down the hill, and at the very bottom of the plant he showed me where his tiny flotation system would go. It was essentially a human-sized box with a rotating internal motor, a miniature version of the one I had seen in the Siglo XX plant. Next to the box was a plastic bag full of xanthate, the powdered chemical used in flotation systems to separate tin from pyrite, and next to that were cans of kerosene and sulfuric acid, ready for use. "Just like the one Patiño used," Mateo grinned. Mateo was an avid fan of Patiño's; in addition to his work as a miner, a metallurgist, and a lawyer, he was also in the process of writing a history of tin mining that would prove that Patiño had been an economic visionary whose legacy had been destroyed by COMIBOL. This is hardly a popular opinion among Bolivian miners (or most Bolivians, for that matter). Mateo's love of mining and faith in its power—particularly its technologically driven power—was informed not only by his surroundings but also by his personal family history, and specifically by the training he had received from his father. His faith was strong enough to carry a significant debt: each of these ancient shaking tables had cost him US$10,000, despite their antique status. He was confident that he could transform what others saw as ruined machines and waste rocks into value for himself and his town.

Part of this transformation process, in Mateo's view, would eventually entail a complete overhaul of the region's extractive architecture. He elaborated on this vision back in the Sink and Float Plant when he and I stopped in the main office to visit with Placido, a cooperative miner who had been appointed to guard and maintain the Sink and Float Plant. A talkative older man in a blue coverall, Placido explained that the Sink and Float Plant technically still belonged to COMIBOL, and the Siglo XX cooperative was slowly paying for the machinery in percentages that were deducted from every sale of concentrated ore to a commercializer or smelter. Placido reflected that it was difficult to justify this expense when the machines were already so out-of-date, and Mateo agreed. He looked at me as he said, "Before, it was cutting-edge technology, but now it's just junk. In Peru they have more advanced technology, people that manage machinery from their computers, open-pit mines that spiral up like a snail, but not here." His words stayed with me because of the disarming way the image of a snail naturalized the pit and even rendered it an aesthetically pleasing use of land.

Placido nodded to corroborate Mateo's assessment of Peruvian technology and informed us that a Peruvian engineer had recently visited the mine. This engineer, whose nationality seemed to validate his expertise, had been confident there was still plenty of mineral ore left in the mountain, despite the century of mining that had taken place there. The hidden geological wealth was at the deepest depths of the mountain, where huge tin veins remained untapped. "It's the mother vein down there, all right," Placido said with a sigh. "But we would need better water pumps to get at it!"

In response, Mateo expressed his disappointment that neither COMIBOL nor the cooperatives had invested in technology the way Patiño had: "We've regressed to the hammer and chisel. We've regressed one hundred years. . . . Before, the passageways and canals were clean, but now you can barely walk down there. We don't have the trolleys and the pulleys, the technologies necessary to clean it out" (Llallagua, February 9, 2017). Alongside Mateo's nostalgia is the hope that new machinery and new expertise would restore the miners to agents of progress. In line with Lauren Berlant's (2011) notion of "cruel optimism," such sociotechnological nostalgia has profoundly negative consequences for miners' bodies, environments, and political lives.

Other miners with whom I spoke were more conscious of these conse-quences and often less invested in replicating the past than in reworking it. This was particularly true for many of the women miners, who shared the technological optimism of their male comrades but had been system-atically excluded from this technology in the past. Many *palliris* spend hours each day traversing the paths that snake through the *desmontes*, usually dressed in full skirts and plastic sandals that offer minimal grip in the case of a misstep. Curlicues of cigarette smoke rise from their work areas—essentially shallow bowls dug into the *desmontes*—which are otherwise invisible from the road. Relying exclusively on hammers and chisels, these *palliris* are some of the most artisanal cooperative min-ers in the region. Further down the hill, where the *desmontes* give way to flat surfaces, other *palliris* work with tin sands that have already been processed at least once. Blackened by the slow sun-induced burn of the acid used in the original concentration process, these sands still contain enough tin to warrant their reprocessing in the women's artisanal con-centration plants, known as *budles* (see chapter 4).

I was first introduced to these women in 2014 by Roberta, the repre-sentative of women miners in the regional federation of Norte Potosí at the time. Roberta, a tall, thin woman in her fifties who flashed frequent smiles of gold-ringed teeth, generously toured me through the edges of town, where *palliris* usually work. She was dressed in mourning because her husband, also a surface miner, had died the year prior. I imagined she must be roasting in her black sweater under the midday sun, but she made no reference to the heat (see figure 5.6). As she walked, she kept one eye on the ground, occasionally stooping to pick up a rock to inspect it for tin. She spit on the rocks she collected, using her saliva to rub away the dust and "make the tin reveal itself." She murmured continuous dis-appointment at how little tin was left in the *desmontes*, pocketing only a few of the rocks along the way.

What Roberta was most eager to show me, however, was not the *des-montes* or the *budles* but a partially completed processing plant that would one day be the exclusive property of women miners. The basic frame of the plant was already there, and all the shaking tables were just outside, ready to go in. Roberta told me excitedly that it was going to be almost as big as the Sink and Float Plant but different in that it would include a meeting hall for women and a daycare for children. The financ-ing had initially come from APEMIN II, a development program that ran from 2004 to 2010 that was designed to support sustainable economic

FIGURE 5.6. Walking through the *desmontes*. Llallagua, July 21, 2014. Photo by author.

development in "poor mining areas" in the departments of Oruro and Potosí. Funded by the European Union (70 percent) and the Bolivian government (30 percent), APEMIN II was the follow-up to a previous collaboration between the European Union and Bolivia (now known as APEMIN I); the main goal of both these collaborations was to prevent unemployed miners and campesinos from migrating to areas in which they could cultivate coca (Oblasser and Chaparro Avila 2008, 42). When APEMIN II ended in 2010, a new, fully Bolivian project called EMPLEOMIN was created to finish pending projects (*La Patria* 2010).[3]

Yet this plant was never finished. When I returned to Llallagua in 2016, the new regional representative of women miners, a shorter, rounder woman named Amalia, explained that the funds had been mishandled. Roberta had paid a company that never completed the project, and now they were short the money necessary to electrify the operation. Throughout the next year, I checked in with Amalia occasionally on how the plant was developing, and there always seemed to be a new problem. Then, in the last month before Amalia was supposed to step down from her position, all the most expensive machinery was stolen from inside the plant. Amalia and her adult son also vanished, leaving everyone to assume that they had stolen the equipment to sell in the city. Before she left, Amalia had lived alone, save a handful of cats, in a single room where she often invited me to drink tea boiled on a hot plate. I mourned her disappearance as much as the incriminating circumstances.

Despite its sorry end, that this plant was being built *at all* illustrated the extent to which even women miners were framing technology as the solution to social and economic problems. Those who disparage how cooperative miners "walk backward like crabs" often point to the persistence and expansion of artisanal mining equipment in the region, which is imagined as a regression from the industrial grandeur of the previous era—not to mention a feminized form of mining. Counterintuitively, artisanal miners agreed with this assessment. For Roberta, the processing plant was going to be a step forward, a way to move from artisanal to machinic processes. Yet, unlike Mateo's technological imaginary, Roberta's step forward was also about gender equity. In addition to greater potential earnings, the symbolic capital corresponding with the advanced technology would have translated into more political influence for women. Moreover, since the plant would have been capable of extracting value from the low-grade ore of the *desmontes*, it would also have transformed the *palliris'* surface-level work areas into much more valuable real estate. This sociotechnological imaginary was specifically about turning waste into value, and the "waste" in question included both ancient heaps of discarded rocks and poverty-stricken *palliris*.

Palliris and other surface workers, although they work with great skill and impressively engineered "artisanal" systems, share with their *compañeros* the belief that large-scale machinery is inherently better. Yet by sweeping the valley clean of artisanal technology and replacing it with symbols of economic prosperity, Roberta also imagined elevating women miners to the same political and economic status as their male companions. Her sociotechnical nostalgia and industrial imaginary were thus shaped not just by the material reality of their surroundings but also by *palliris'* precarious positions as women miners and surface laborers.

UNDERGROUND INFRASTRUCTURE: INSIDE THE MOUNTAIN

The mining engineer Raúl was officially retired by the time I met him, but he was still teaching the occasional class in the mining engineering department of UNSXX. Seated in his kitchen, where the walls were decorated with cheerful images of baby animals, Raúl took it on himself to explain to me exactly how Patiño had extracted tin from the Juan del Valle Mountain. I told him that I had already interviewed several geologists, but Raúl was profoundly dismissive of these earth scientists, whom he claimed never spent any time underground—unlike mining engineers, who "lived in the rock." Nor did he think much of the cooperative miners, whom he described as "illiterate plunderers" (*saqueadores analfabetos*). His grudging admiration was reserved for Patiño, and Raúl wanted me to understand why this was the case.

After Patiño had already exploited the most lucrative veins in the Juan del Valle Mountain, Raúl explained, his profit margins started dipping precipitously. The mine was still full of mineral, but it was strung around like rosary beads instead of forming solid veins of ore. Because exploiting rosary beads individually was slow and expensive, in the late 1930s Patiño started going around "knocking on doors" to find engineers in Europe and Asia interested in devising a new technique adequate to the challenges Patiño was facing in Bolivia. Eventually, a group of Japanese engineers came up with "block caving," a method that—Raúl claimed—had never been tried anywhere else. Raúl pulled out the textbook he was using to teach his undergraduate course and flipped through the pages until he found a diagram. Essentially, block caving involves dividing the mine into subterranean cubes and drilling a series of funnels under each cube. If all goes according to plan, when the cube is detonated, the rocks fall through the funnels, down a chute, and into the transport trolleys waiting below. "Bolivia was the pioneer in the use of block caving in the world, which was later used in Peru and Chile," Raúl reminisced, showing me diagrams of similar techniques used in other mountains. "It was Patiño who brought that technology. He was a genius, and that's why he was one of the seven millionaires in the world at the time." The legacy of this "genius" is evident inside and outside the mine. Raúl described block caving as taking "spoonfuls" out of the mountain, and this was indeed what it looked like from the outside. While formerly detonated blocks appear underground as empty galleries—ballroom-sized spaces filled with jagged boulders—on the outside of the mountain it looked like someone had spooned out the middle of a cake and left behind the frosting to slowly collapse under its own unsupported weight.

I toured many of these empty galleries with Hilario, the older miner introduced in chapter 4 who drives the subterranean trolley cart. As we picked our way slowly underground, he was fond of having me guess what the spaces had been used for in the past, a game that I never won. For instance, we once paused next to a cavernous room heaped high with plastic water bottles that had once been an underground office, where timecards were stamped and secretaries tapped on typewriters. There had even been a phone line so that underground managers did not have to walk all the way to the mine's exit to communicate with their superiors. Another cave, located along the same tunnel, had been a jail for *jukus* (ore thieves) and workers who showed up drunk. Its ceiling had since caved in, snapping the wooden framework of the room like toothpicks. Splinters of soggy wood were all that remained of the jail, a fact that seemed to physically pain Hilario. He leaned on the side of the tunnel and shook his head, gesturing disgustedly to all the plastic bottles and rocks strewn around where jail cells had once been. "It used to be clean, modern, efficient," he said, "but now these *jovenes* [youngsters] don't have any respect." Respect, he underscored repeatedly, was what was missing among young cooperative miners: respect for the mine, the work, and the past.

Yet my impression of the young miners was somewhat different from Hilario's. It seemed to me that even if they were not caring for the mine in the old way, these miners nevertheless respected the values that the ancient infrastructure represented. For Hilario, the office spaces and jails and telephone lines were all symbols of industrial progress, and their meanings had been tarnished as they fell into disrepair. But for young miners, the commitment to progress was materialized elsewhere. For example, they often wanted to pose for photos next to air compressors, large cylindrical machines that send air to miners' pneumatic drills. These air compressors, which are usually owned by a single *cuadrilla* (work crew) or individual, are lovingly decorated with *mistura* and *serpentinas* (confetti and paper garlands) during every major celebration, their shiny ovular exteriors wiped free of any dirt or debris. In other words, the promise of sociotechnical progress may have been materialized in different objects for different generations of miners, but the allure of a better future was consistent throughout the subterranean space.

This was also true for Esteban, a mining leader whose story I chart in detail in the following chapter. Esteban focused on showing me underground electrical generators, which he encouraged me to photograph until I protested that I had seen enough. He then took me to see two operational elevators and one more that had been the cause of a devastating accident several years prior. The cord had snapped, sending half a dozen cooperative miners to their deaths. The elevator operator, who had been responsible for ensuring that none of the cables showed signs of wear, had been imprisoned for carelessness, and the elevator had not been resuscitated. The two working elevators, however, did not feel significantly safer. The miners call them *jaulas* (cages), and this is an accurate description of how they feel. These rectangular boxes are open on two sides so that miners can easily stream in and out, but the lack of doors means that the rockface is stomach-churningly visible as it hurtles past. The *jaula* also lacks buttons of any variety, so miners communicate with the operator by tapping live wires together, Morse code style. When they get no response, they assume the operator has left for the day, and they climb the elevator's external frame as if it were a ladder. Esteban was excessively irritated when this happened to us, in part because he had been so eager to show me a working *jaula*.

But even if they had been working, we would still have had to climb down the last fifty meters, since these elevators only carry miners to level 750, or 750 meters down from the mountain's peak. Below that, the chute has narrowed too much for the miners to be confident that the elevator can pass. The miners themselves cannot climb much more than fifty meters lower than the elevator since their water pumps cannot keep up beyond that point. Yet the miners keep trying, as it is in the most inaccessible depths that they are convinced that they can strike it rich. Rumors of new deposits submerged beneath the water circulate endlessly, even though only a handful of formal geological tests have been done in the mountain since COMIBOL folded in 1985. Traveling geologists arrive regularly in Llallagua, like prophets in 4X4s, promising cooperative miners that the lower levels of the mountain contain unbelievable wealth, but these claims remain unverified because there is too much water for exploration.

Because of this uncertainty, *cooperativistas* often pass the time by fantasizing about the depths that could be reached with newer technology. Sitting in *pauwiches* (storage rooms) with groups of miners in the morning, I was regularly called on to comment on the possibility of turning their mountain into a pit like the ones they had seen pictured in Peru and Canada. I was taken aback that they would even want such a mine, particularly given Marisol de la Cadena's (2010) contention that there is a qualitative difference between shaft mining and open-pit mining for Andean peoples. It was also difficult to imagine how an underground world so full of machinery, clothing, and caverns of discarded plastic bottles could be dismantled. I usually thought only to comment on the small number of jobs offered by open-pit mines, to which the miners would respond with disappointed headshaking. Only when going through my field notes much later did it occur to me to wonder at the number of miners who tied visions of technological salvation to invisible geological wealth. Miners found hope in this articulation, which I inadvertently unhooked by pointing to the potentially negative consequences of their sociotechnical imaginaries.

Between the galleries that were blasted open by block caving and the skeletal framework of tunnels, shafts, and elevators that was built in the early twentieth century, the mine's internal spaces materialize a history of progressively complex extraction. Within and around these spaces, generations of miners have also left their mark in discarded clothing, food containers, and spaces carved expressly for the purpose of relaxing. Through their daily interactions with the objects and the space itself, miners experience nostalgia for an imagined future past that bleeds directly into an imagined future present.

LIFE CYCLE OF METAL

There are two annual holidays universally celebrated by miners, cooperative and salaried alike: Carnival and August 1. While Carnival is widely observed in association with the beginning of Lent, the first day of August is an important day in the agrarian calendar that has been transposed into the mining world. August is typically the driest month of the year, and Andean farmers welcome it with celebrations that help ensure crop survival. In the mines both Carnival and August 1 are similarly ritualized by spilling either llama's blood or (much more commonly) alcohol in subterranean spaces. "Que se mejoren las estructuras" (Here's to improved [mineral] structures), say the miners as they drip beer, wine, or *trago* (spirits) into the corners of their work areas.

But it is not just the rocks that receive this special treatment. Semi-trucks, elevators, abandoned buildings, trolleys, and ore mills are all festooned with soggy paper garlands, and I have also seen miners flicking beer at radios and cable TV hookups. These blessings are demonstrations of respect for industrial ruins that remind residents of a time when the towns' mining sector was the motor of national history. As illustrated in the mural at the beginning of this chapter, Llallagua-Uncía prides itself on having produced leaders whose actions reverberated at national and even global scales: union leaders, Trotskyist historians, and revolutionaries. Although these leaders are commemorated by statues and murals, the context of their emergence is also unofficially memorialized in the towns' many relics. Cooperative miners' day-to-day labors are motivated by twentieth-century ideals of progress and the still-circulating dream of a past possible future, lodged deep underground.

Perhaps the clearest example of this subterranean promise came when Mateo and I exited the Sink and Float Plant after the tour. I was asking him why the state had not simply sold all this machinery in the 1980s, particularly if it was still valuable enough now to warrant a military guard. Mateo averred that the sale of machinery was not allowed. Instead, any unwanted equipment had to be buried in the ground. He recounted a tale about a past FERECOMINORPO president who had discovered a powerful ore mill buried in the earth near the *desmontes* and sold it to raise money for the federation. When his actions were discovered, this man had been imprisoned for the theft of state property, but this did not stop others from digging in hopes of making a similar discovery. Literal machines buried in the earth, like an industrial treasure chest. I laughed out loud at the ghoulish life cycle: from the earth the metal comes, and to the earth it is returned, with the hope that it could be made to rise, zombielike, to work again.

Sociotechnical nostalgia, materialized in machinery and waste spread across postindustrial landscapes, whispers of past grandeur and future wealth in ways that shape how miners act in the present. In the next chapter, I explore these actions as they unfold across the contentious boundary between the state and mining cooperatives, a boundary marked by patronage, punishment, and corpses.

6

GEOLOGY OF *PATRIA*:
PATRIMONY, PATRONAGE, VIOLENCE

As in most of Bolivia, Carnival is a big deal in Llallagua. By late afternoon on a Friday in February 2016, the first day of the weeklong festival, the downtown was littered with confetti, paper streamers, and plastic cups that had been ground into the mud by thousands of dancing feet. A continuous light drizzle—February is the end of the rainy season—dampened no one's spirits as parades of dancers, brass bands, and groups playing traditional music processed through the streets. School-age children peeked out from behind corners to pelt unsuspecting passersby with water balloons, while adults greeted friends and family members by wrapping their necks in streamers, sprinkling confetti on their hair, and toasting the year ahead.

For miners, however, the real celebration begins on the Saturday of Carnival, when everyone descends into the mine to decorate and bless one another's subterranean staging grounds (*pauwiches*) and work areas (*parajes*). Pouring small amounts of alcohol or (for the more ambitious) llama's blood into the corners of a *paraje* ensures both the appearance of valuable minerals and the protection of miners from accidents. Miners each decorate their own *parajes* first and then travel throughout the mine paying visits to one another and to the clay Tíos, the devil figures that are typically formed with open mouths to receive gifts of lit cigarettes, fresh coca, and full bottles of wine and beer. Like the miners themselves, their necks are wreathed in streamers, and confetti is strewn across their heads and feet.

Don Esteban, former president of the regional mining cooperative office FERECOMINORPO and current plurinominal deputy for Norte Potosí in Bolivia's Plurinational State, invited me to join him for the underground festivities. He had been living in La Paz since Evo had been reelected in October 2014—a success that catapulted all the MAS's plurinominal candidates (chosen by the party instead of by popular election) into parliament—but he had returned home for the holidays. I accepted his invitation with excitement and took a shared taxi to meet him at the Cancañiri mine shaft on Saturday morning. I was sandwiched between miners dressed for a subterranean party, plastic bags bursting with decorations overflowing their laps and shoved down near our feet. The taxi parked in the middle of the miners' market, where hundreds of men milled about in rubber boots, sweatshirts, and helmets. Along with the usual goods for sale (coca leaves, miniature bottles of alcohol, cigarettes, and snacks), merchants were selling paper streamers, confetti, shallow terracotta plates filled with offerings, white and pink balls of powdered sugar with fennel at their hearts, and white sugar squares with images printed on them representing different hopes for the year ahead: wealth, education, business success, housing, children. Once we were underground, things were even livelier. The elevator operator, already a round man, was completely wreathed in streamers and dancing by himself to Peruvian *chichas*, songs that blend tropical and Andean musical traditions with psychedelic and surf rock. Everyone was laughing and, over the tinny sounds of music blasted from cell phones, making plans to meet at different Tíos after they had blessed their respective work areas. Fresh layers of streamers and coca leaves had been added to every corner of the mine, turning the interior space of the mountain momentarily colorful. Esteban, his son, and I decorated their *pauwiche* with streamers and delivered a terracotta plate piled high with sugary offerings into a horizontal crack in the wall.

Settling down to share a bag of coca leaves, Esteban explained that it was important for him to return regularly to his *pauwiche* and *paraje* to "maintain the custom" of being underground. From what I could observe, this custom was about maintaining collective ties with the cooperative as much as maintaining his individual practice of subterranean labor. Esteban had won his position as plurinominal deputy with the support of many other cooperative miners and their families, and he needed to reassure them that he had not forgotten where he came from, or to whom he owed allegiance. Therefore, instead of spending much time in his *paraje*, we spent most of the day underground traveling from *pauwiche* to *pauwiche*, from Tío to Tío, to speak with as many miners as possible.

In addition to showing his face and reminding people of his existence, Esteban was also taking the opportunity to campaign. Hardly more than a year into his third term, Evo had announced an intention to run for president a fourth time in 2019, and Esteban had been tasked with drumming up support for this announcement. It was more controversial than a regular election because Evo was constitutionally allowed to hold the position only twice and had already managed to wrangle a third term on the grounds that the constitution had been rewritten during his first few years in office, which meant that his first term should not count against him. Now Evo was holding a general referendum that, if the *sí* (yes) vote won, would allow him to amend the constitution to make a fourth term possible. The referendum was scheduled for February 21, just two weeks after Carnival, and Esteban's underground mission was to convince the miners to vote yes. As we moved around, accompanying different *cuadrillas* (work crews) in their underground blessing, Esteban made a point of reminding everyone of what "compañero Evo" had done for them. He directed their attention to electrical generators and air compressors that had been provided through the Evo Cumple, Bolivia Cambia (Evo Delivers, Bolivia Changes) program, and he inquired as to whether any of them might take advantage of the early pension plans that Evo had made available to those with physically challenging occupations, like mining. Most of the men nodded as Esteban talked, seeming to agree with him that Evo had to stay in power for the good of the cooperatives. Campaigning had never looked so easy.

Esteban received the most pushback on exiting the mine at the very end of the day. After nearly seven hours of subterranean celebration, he and I staggered back into the miners' market to share a few beers with Pocha, a woman who works in the miners' canteen and who had served us peanut soup early that morning. By this point in the day, Pocha had sold out of soup and was busy celebrating Carnival while she watched over her husband, who, having done his celebrating earlier, was snoring gently in her lap (she laughed that he was a good husband because he always passed out where she could keep an eye on him). Sidestepping Esteban's inquiries about her stance on the referendum, Pocha instead took his presence as an opportunity to gather details about a scandal that had just broken in the national news. Evidence had emerged indicating that from 2006 to 2007 Evo had maintained a relationship with Gabriela Zapata, a then-twenty-year-old *q'ara* (white) Bolivian woman who had been hired by the Chinese company CAMC to lobby for company interests in the newly consolidating Plurinational State. Although the story kept shifting, reporters had recently been claiming that the couple had had a child, whom Evo claimed died shortly after childbirth, while Zapata insisted the child was alive and living abroad.[1] Cradling her own, comparatively well-behaved partner's head, Pocha pressed Esteban on the details. Wasn't it just *wrong* of Evo to have had the affair, and especially to have abandoned the woman and child afterward? Esteban deflected: men are weak, Evo was single, the company used the young woman to tempt him, and in any event Evo's personal life should not count against him in the referendum. Esteban nodded toward the canteen stalls where Pocha worked, which had recently been rebuilt with government money, and asserted that Evo's dedicated service to the *pueblo* (people) far outweighed any individual harms. Pocha, however, looked unconvinced.

The Zapata scandal raised the ire of many Bolivians, not just Pocha. In addition to the familial concerns, journalists had uncovered seven major contracts signed between CAMC and the Bolivian state since 2006, six of which were not open to bidding, and all of which dealt directly or indirectly with resource extraction. In a context of increasing concern about Chinese influence in Bolivia, particularly within the extractive sector, this story reignited a debate about whether the Bolivian economy had really transformed at all under Evo or whether it had just traded the influence of one imperial power (the United States) for another (China). Critical analyses in this vein focused on the Zapata scandal as evidence of Evo's corrupt leadership style, which appeared to echo Bolivia's long history of national elite collusion with imperial powers.[2] Much later, I reflected that the two concerns raised by the Zapata scandal, the familial and the imperial, were not as disconnected as they appeared. Both were narratively organized as stories of betrayed patrilineal inheritance: in the former, a child is denied his rightful inheritance because of his father's irresponsible behavior, while in the latter, the Bolivian nation is denied its rightful inheritance (natural resources) through illicit deals struck by an irresponsible leader. The connective tissue between the two narratives was Evo himself, figured in both stories as a failed father.

When Evo lost the February referendum, 48.7 percent to 51.3 percent, the defeat was largely credited to two scandals, one of which was the Zapata affair and the second of which was a separate, much longer-running conflict that similarly involved an apparently corrupt transfer of wealth derived from natural resources.[3] In late 2015 news had broken that hundreds of projects financed by FONDIOC (Fondo de Desarrollo para los Pueblos Indígenas Originarios y Comunidades Campesinas; Fund for the Development of Original Indigenous Peoples and Campesino Communities) had never been completed. This scandal involved the detention of not only the government officials involved with the fund but also many Indigenous leaders, who were accused of having siphoned off the money for personal use. Many activists claimed that this scandal was orchestrated by the state to imprison Indigenous leaders who had been critical of the MAS (Evo's party) and install puppet leaders, but others took it as evidence that Evo had long been transferring natural resource wealth to corrupt Indigenous leaders in exchange for political support. Taken together, these scandals were challenges to three dimensions of patria (nationhood): patrilineal descent (Evo as father), patrimony (natural resources as collective inheritance), and patronage (politically motivated gifts). At the heart of all of these are questions about nature, and more particularly subterranean resources. Who owns these resources, and whom should they benefit?

Earlier chapters of this book explored how the multiple material histories of Bolivia's subterranean have been shaped by geological, colonial, nationalist, and extractivist histories and how this subterranean, in turn, has shaped socially differentiated collectives and bodies. In this final chapter, I am concerned with how the subterranean, as historical matter, shapes contemporary political dynamics at the national level, a process that I trace along two interconnected pathways. First, an abstracted sense of the subterranean as national inheritance (patrimony) undergirds dynamics of political patronage and political violence. Twins of one another, patronage and violence are rooted in colonial histories of resource extraction, and they are maintained in contemporary political arenas through an ongoing construction of the subterranean as a shared treasure chest, the economic as well as physical foundation of nationhood. In other words, if chapter 1 historicized the materialization of the Bolivian subterranean as patrimony, this chapter explores how this history is linked to social practices of patronage and violence. Second, when people who have been individually and collectively shaped by intra-actions with subterranean matters—a process that I explored, from various angles, in chapters 2–5—move into national-level politics, they bring with them subjectivities that were forged underground. The establishment of the Plurinational State in 2009 created a new host of pathways for previously excluded people to take on leadership roles within or alongside state entities. Cooperative miners, who quite literally retain in their flesh geological matters imbued with colonial and industrial meanings, were suddenly making political and economic decisions on behalf of a newly constituted Plurinational State. Thus, although most of this chapter is staged in arenas that appear purely social, my goal is to draw attention back to the subterranean matters that have shaped the dynamics of patrimony, patronage, and violence, as well as the participants of these dynamics. Geological matters, as historicized throughout this book, have left their mark not only on flesh and bone but also on the hallowed halls of political and economic decision-making. The subsoil is always already present in economic, political, and social forms.

My entry point into this discussion is the interwoven stories of two miners from Llallagua, both leaders within the cooperative community, both either entering or attempting to enter the institutions of the state. One of these is Don Esteban, who moved from the mines to parliament in 2014, a position that made him a sometimes-unwilling mediator between the state and the mining cooperatives. The other is Don Alfonso, introduced in chapter 2, who was president of Norte Potosí's regional federation of mining cooperatives (FERECOMINORPO) from 2015 to 2017. Over the course of this period, Alfonso attempted to jump from his regional leadership position to a central government ministry. The key political eruption around which both Esteban's and Alfonso's stories turn was the murder of Rodolfo Illanes at the hands of cooperative miners in late August 2016, an event that I began to describe in the introduction and to which I return in more detail now. The myriad of political tremors that reverberated in the mining districts before and after this event demonstrate how geological matters become an object of struggle through which state and society were mutually constituted.

Under the leadership of Evo Morales, the Bolivian government aimed explicitly to redistribute both economic wealth and political power. Among the mining cooperatives of Norte Potosí, these two forms of re-distribution sometimes seemed to run counter to one another. On the one hand, resource nationalist policies—that is, policies designed to redistribute wealth derived from resource extraction throughout the country—enriched local mining cooperatives and subsidized their ac-tivities. On the other hand, cooperative miners have frequently used their increased political power to oppose taxation that would have re-sulted in the redistribution of their own mining profits. This appar-ently contradictory position is rooted in a complex set of identifications with the subsoil as both national patrimony (i.e., belonging to all Bo-livians) and local patrimony (i.e., belonging just to residents, or even just to those descended from miners). As Elizabeth Emma Ferry (2002, 2005) has detailed in her study of silver-mining cooperatives in Mex-ico, it is not unusual for these differently scaled sets of patrimony to co-exist, particularly in the aftermath of industrial extraction. In Bolivia the commitment to protecting local patrimony deepened in response to the prior experience of neoliberal central state abandonment. Although the MAS government was generally responsive to the needs of the coun-try's highland mining districts, the 1985 gutting of the tin-mining sec-tor still weighs too heavily for most miners to assume that government support will persist indefinitely. And, indeed, the events of 2016 re-vealed the limits of this support.

At its core, the model of economic reform undertaken by the MAS gov-ernment was based on a commitment to subterranean redistribution—or, more precisely, a commitment to redistributing the wealth derived from subterranean resource rents. As many scholars have noted, Evo's rise to power coincided with the commodity boom, and one of his first actions on taking office in 2006 was to nationalize the country's natural gas sector.[4] The process of redistributing gas rents, however, had already started the year before, when the interim government of Carlos Mesa in-troduced a new tax on natural gas and petroleum known as the IDH (Im-puesto Directo a los Hidrocarburos; Direct Tax on Hydrocarbons). The IDH is a nationalist tax in that it aims to redistribute rents derived from subterranean resources in one region throughout the rest of the country.

It benefits all nine departmental governments, 340 municipal governments, the national pension plan, the social grant programs for qualifying individuals (such as mothers and students), the national treasury, and FONDIOC (Aresti 2016). Starting in 2007, it also financed (with assistance from the Venezuelan government) the Evo Cumple, Bolivia Cambia program, which funded the construction of *obras* (public works) such as highways, sports arenas, city halls, and hospitals, primarily in underserved areas. At 32 percent levied on the gross value of hydrocarbon production, the IDH had already transformed the fiscal makeup of the central state before Evo ever set foot in the presidential office. The IDH and the natural gas sector's subsequent nationalization both align with the widely held conviction that natural gas is the inheritance of all Bolivians, or a shared patrimony. As Fernando Coronil (1997) describes for oil rents in Venezuela, the sudden infusion of natural gas rents lent the MAS government a "magical" quality, as if it were capable of literally pulling money from the earth. For many people, the infrastructure funded by Evo Cumple, Bolivia Cambia materialized a subterranean patrimony that they themselves had never laid eyes on.

But not all Bolivians were so wild about Evo's redistributive strategy. New and traditional elites who had benefited from the previous era's neoliberal emphasis on political devolution were particularly resistant to change. As scholars have observed in other contexts around the world, neoliberalism resulted in a proliferation of local state and extrastate involvement in resource governance, and these fortified authorities now expect to be at the helm of any new extractive or industrial project taking place within their territories (Coyle Rosen 2020; Luning and Pijpers 2017). Especially at the beginning of his time in office, Evo's government met the greatest resistance in the departments of Tarija, Santa Cruz, Beni, and Pando, which collectively described themselves as *la media luna* (the half-moon) because of the shape they form along the country's eastern border. Almost all the country's natural gas fields are located within Tarija and Santa Cruz, especially within the Gran Chaco region of Tarija. Traditional elites in these two departments resisted the IDH, Evo's election, and the nationalization of the natural gas sector because they believed that the benefits of resource extraction should accrue to the regions where the resources are located. Such resource regionalist concerns were placated with the maintenance of a royalty system in which 11 percent of the gross sales value is distributed back to hydrocarbon-producing departments (and 1 percent is reserved for the two poorest departments, Pando and Beni, which tend to be political allies of Santa Cruz and Tarija despite not having any natural gas).[5] Bolivia thus maintains something of a dual system of natural gas rent distribution, one that aligns with resource nationalism and one that attends to regionalist concerns. The subsoil is redistributed, but not equally.

This is not dissimilar to what Esteban and Alfonso sought for Norte Potosí. As leaders of their mining cooperatives and communities, Esteban and Alfonso both worked hard to direct the redistributive benefits of natural gas extraction toward their constituents, but they also sought to protect the system in which royalties paid by cooperative miners are returned to their departments (85 percent) and municipalities (15 percent) of origin—the central state receives no benefit from cooperative mining and therefore cannot redistribute it. The MAS was amenable to this apparently contradictory agenda so long as cooperative miners could guarantee regional success for the MAS in elections at all scales. This agreement was not entirely novel, since the MAS has always relied on the coordinated efforts of social organizations—such as workers' unions, neighborhood associations, and cooperatives—to win hearts, minds, and elections; such active grassroots engagement is the reason the MAS government has been described as a "social movement state." When Evo came to power, however, he began bestowing development-oriented "gifts" such as electrical generators, soccer fields, semitrucks, and sports arenas on supportive municipalities and federations. All these gifts were made possible by the influx of natural gas rents, prompting Bret Gustafson to note that "the gaseous state has now overtaken the social movement state" (2020, 174). In Llallagua-Uncía, as elsewhere, the arrival of gaseous gifts was accepted as reasonable compensation for the hard work of participating in a social movement. The municipal governments, as well as the regional federation of mining cooperatives, all benefited from the flow of gas rents, which arrived through the coordinated efforts of leaders like Esteban and Alfonso.

This developmentalist dynamic represented a major shift from the neoliberal era, in which aid organizations such as Oxfam, World Vision, and various foreign development agencies were the most significant sources of financial assistance for rural projects in a context of relatively reduced state capacities. Two changes provoked the rollback of these agencies' operations. First, in response to both Bolivia's economic growth and global financial contraction since 2008, many aid organizations began voluntarily reducing their operations in Bolivia. Second, the MAS government began actively expelling and discrediting the international development community. Evo personally expelled USAID (United States Agency for International Development) in 2013, and in 2015 Vice President Álvaro García Linera mounted a campaign against Bolivian NGOs that relied on external funding and published works critical of the government (Achtenberg 2015). While explicitly intended to restrict the imperial reach of Europe and North America, these moves also succeeded in reducing any political opposition that might emerge from rural and cash-strapped collectives. By 2016–17 the state—and particularly its infrastructural arm, Evo Cumple, Bolivia Cambia—dominated the rural development scene; to critique the MAS would have been to turn off the financial tap. For instance, the highland Indigenous federation CONAMAQ began distancing itself from the MAS after the 2011 conflict over TIPNIS (Isiboro Secure Indigenous Territory and National Park), and by 2016 the federation was struggling to pay rent for its office spaces, let alone host major meetings.[6] The state was not interested in funding its own opposition, and the state had become the only game in town.

As local leaders, Esteban and Alfonso had to adjust to the new rulebook to ensure that economic redistribution policies continued to benefit Norte Potosí. As individuals, however, they also sought to make use of openings created by political redistribution policies to advance their own careers. In the service of popular democratization, the 2009 constitution introduced a new system of electoral representation that combines direct and proportional processes. The Plurinational Legislative Assembly consists of a lower house (Chamber of Deputies, 130 seats) and an upper house (Senate, 36 seats); of the deputies, 70 are elected to represent single-member electoral districts (*diputados uninominales*; uninominal deputies), 7 are selected by Indigenous districts, and 60 are elected by proportional representation from party lists on a departmental basis (*diputados plurinominales*; plurinominal deputies). For people in rural areas, this system has translated into more opportunities to enter national politics, especially since the MAS tended to choose plurinominal candidates who had already been leaders of political, economic, or social institutions in rural communities (Galindo 2017). The plurinational assembly, unlike perhaps any other parliament in the world, was suddenly filled with leaders of peasant unions, workers' unions, workers' cooperatives, Indigenous federations, women's guilds, and other community-led institutions. People whose formal political activities had previously been constrained to either voting or protesting were now involved in crafting the legal architecture of the state.[7]

When I first met Esteban in July 2014, he had just been selected as the plurinominal candidate for the MAS in Norte Potosí, which meant that he would automatically join the Chamber of Deputies once Evo was reelected as president (which happened in October of that year). At the time, I was impressed by how well Esteban met my expectations of a local leader: although he dressed similarly to other off-duty cooperative miners, in a polyester tracksuit and a baseball cap, his glasses and thick mustache gave him an intellectual look, and he strode through the town's plaza with the purpose of a local celebrity. Although he had exchanged his tracksuit for a long navy overcoat and flat cap when he moved to La Paz, he made an ongoing effort to assure everyone in Llallagua that he was still part of the *pueblo*. Alfonso, while not yet as influential as Esteban, similarly seemed to be on his way to achieving what Esteban already had. A tall, lean man with tired eyes and an easy smile, Alfonso had been the president of FERECOMINORPO for nearly a year when I met him in early 2016. He and others around him assumed that by the end of his two-year term, he would follow in the footsteps of previous FERECOMINORPO presidents, who included not only Norte Potosí's plurinominal deputy (Esteban) but also Norte Potosí's senator and Llallagua's mayor. All Alfonso had to do was avoid losing the support of either the miners or the government for one more year, and he would be set for at least several more years of political service. The referendum of February 2016 therefore mattered to Alfonso's professional future, perhaps even more than it did for Esteban.

Yet despite Esteban's assurances to the miners, Pocha, and everyone else, Evo lost the referendum. In Llallagua the immediate impact of this loss was a halt on the construction of a sports arena that had been described to me as a gift for the mining cooperatives from the MAS government (its construction was funded by Evo Cumple, Bolivia Cambia). Alfonso moaned that this was Evo's way of punishing him and his directorate for not having drummed up enough votes. "My own daughter voted 'no' because of the Zapata scandal!" Alfonso lamented when I saw him the week after the referendum. "I told her that what Evo does in his private time is his business, but she voted no anyway! What am I supposed to do if I can't even convince my own daughter?" To make matters worse, Evo's defeat in the referendum also eroded support for Alfonso among cooperative miners. Alfonso was a member of the MAS, and cooperative miners who belonged to other political parties began agitating for a change in FERECOMINORPO leadership to better represent the interests of cooperative miners. While navigating this newly complicated political context, in which it seemed that he could satisfy neither his constituents nor his political party, Alfonso was called on by the mining cooperatives' national-level leaders in FENCOMIN to participate in the August 2016 roadblock.

Roadblocks are the classic, most frequently used form of protest for social collectives across Bolivia, both miners and nonminers alike. People use their bodies, cars, stones, and whatever else they have at hand to occupy public thoroughfares and demand government attention (Bjork-James 2020). There is a geological history to the efficacy of this tactic in Bolivia, even though contemporary roadblocks often have nothing—or at least nothing obvious—to do with the subterranean. At its core, the goal of a roadblock is to increase protesters' leverage by threatening economic stability.[8] Commodities, whether natural resources or manufactured goods, must circulate to realize their value; from a national economic standpoint, the unceasing circulation of "strategic" resources is therefore of the utmost importance. Since colonial times, Bolivia's most strategic resources have all been derived from the subterranean: silver, tin, and natural gas. While natural gas, as a liquid, can be transported via pipelines instead of along roads (a reality that has ignited new sets of political practices in Bolivia and worldwide), silver and tin ores have traveled primarily along railroads and highways.[9] Indeed, the need to move metal ores is the reason that much of this infrastructure was built in the nineteenth and twentieth centuries, and it was often through blocking such movement that miners' unions won significant democratic gains for the nation. Even though today's roadblocks are less specifically concerned with the movement of strategic metals than with the movement of people and goods in general, the roadblock remains highly effective, since Bolivia's comparative lack of highway infrastructure means that a single blockage can result in a major shutdown.

The roadblock that cooperative miners erected in August 2016 was sited in Panduro, along the only road that connects La Paz directly to Oruro. When Alfonso received word that he should mobilize the miners in his federation, he passed the order along to the individual cooperatives, telling each to reserve buses and send at least a quarter of their members to join the protest. As the miners began to congregate in Panduro, FENCOMIN released its ten-point list of demands, which included the right to form partnerships with private companies (without state oversight), expanded work areas, and relaxed environmental regulations. And the miners settled in to wait.

Illanes's murder on August 26 marked a definitive turning point for relations between mining cooperatives and the state, but tensions had been building for a long time. The conflict brought to light the extent to which the MAS administration, despite distinguishing itself in many ways from its neoliberal predecessors, still relied on colonial-era dynamics of patronage and violence to maintain peaceable relations with the mining cooperatives (as well as other social groups or movements). The conflict's aftermath also demonstrated the extent to which these dynamics are operationalized through the subsoil.

GEOLOGICAL GIFTS: *PATRONES* AND PATRONAGE

In late September 2016, only a month after Illanes's death, I joined around three thousand cooperative miners from across the country in a sweaty sports arena on the outskirts of Cochabamba. This national congress had been hastily organized, and the venue had been hard to secure; most municipalities were hesitant to host cooperative miners in their communities so soon after the miners had made international headlines for killing a government official. The atmosphere was tense, and Don Alfonso invited me on the condition that I sat only with miners from Norte Potosí, who had in turn been instructed to look out for me. One of the other miners fashioned me a credential—a stamped piece of paper that certified my connection to the federation—and strung it around my neck like a protective charm.

The major task for this meeting was selecting a new set of leaders for FENCOMIN. Since the entire existing directorate had been thrown in jail, the composition of the new directorate would signal to both the miners and the MAS whether reconciliation or retaliation would be the driving intention of the reconstituted organization. In addition to choosing new leaders, however, the miners also needed to figure out a plan to free more than fifty cooperative miners who had been imprisoned following Illanes's murder. This task promised to be challenging, given that public opinion of cooperative miners had never been lower. There appeared to be multipartisan agreement across major news outlets that cooperative miners were nothing more than plunderers, thieves, and *patrones—*

a word that can be translated as "bosses," "patrons," or "owners"—who deserved any jailtime they received.

As discussed in the introduction, the conflict that had ended with Illanes's death had initially been sparked by a proposed modification to the General Law of Cooperatives (Law 356) that was slated to recognize the right to unionization for workers employed by "cooperative societies." Although they have never been explicitly permitted to do so, mining cooperatives have a reputation for hiring workers—poorly compensated and sometimes children—to labor on their behalf, a system of exploitation that could be threatened by unionization. Because of this labor arrangement, cooperative miners are often referred to pejoratively as *patrones*. Despite the long list of demands that the cooperative miners at the roadblock produced—only one of which had anything to do with the cooperative law or the threat of unionization—journalists reporting on the incident seemed to concur that cooperative miners had taken to the streets because they were more like *patrones* than workers and therefore inherently antiworker in their politics. As a journalist acquaintance of mine in La Paz put it, "Those miners are gangsters [*mafiosos*]. I think they're the most hated group in Bolivia—everyone hates them! *Malditos empresarios con nombre de cooperativas* [Goddamned entrepreneurs calling themselves cooperatives]." He banged his whiskey cup on the table for emphasis.

The figure of the *patrón* and the relational dynamic that it conjures—*patronaje* (patronage)—is key to understanding how cooperative miners operate internally, how they make claims on the state, and why they are so deeply unpopular. In disciplines such as sociology and political science, patronage is a central concept that describes an arrangement in which a political leader (patron) transfers infrastructural, financial, or symbolic powers to their constituents (clients) in exchange for political support, often but not always in the form of votes (Auyero 2000; L. Taylor 2004). But in Latin America patronage is also the direct descendant of *patronaje*, a colonial-era dynamic in which men who owned and operated mines and haciendas (plantation-style farms) were all called *patrones*, a title that implied a position of authority with some reciprocal obligations. In (forced) exchange for hard labor and taxes, the *patrón* was supposed to "care" for Indigenous workers by providing (unwelcome and/or wholly inadequate) "services" such as housing and religious education. Although this relationship was coercive, it also created the only mechanism through which Indigenous peoples could express any form of entitlement. That the root word of *patronaje* is the Latin *pater*, or father, is no etymological accident; the figure of the *patrón* as landlord is closely connected to the figure of the *patrón* as father, in the disciplinary and dependent sense. Indeed, in addition to existing relations of violence, exploitation, and putative care, Indigenous workers often invited their *patrones* into relations of *compadrazgo*, or godparenting, a system of financial redistribution that remains operational across the Andes and Latin America more generally. *Compadres* (literally "joint fathers") were and are expected to assist their godchildren with expenses ranging from school supplies to soccer uniforms to wedding-party provisions, thereby ensuring that the affluent members of a community redistribute their wealth. By the same token, however, this elevation of a colonial authority to the head of a heteronormative family worked to feminize and infantilize Indigenous peoples, or to symbolically figure them as requiring paternal direction. The entitlements that Indigenous peoples won, either in the form of *compadrazgo* or more general elements of colonial *patronaje*, also deepened economic dependencies and came loaded with expectations of loyalty, making it harder to upend existing social hierarchies. Debts, some social and some financial, increasingly bound workers and campesinos to their employers and fictive kin. This combination of physical, economic, and social coercion (violence, debt, and obligation) worked to lock in colonial relations long after the colonial period officially ended.

Contemporary patronage, therefore, is as much about dynamics of corporativist entitlement and dependency inherited from the colonial era as it is about "buying votes" in the modern sociological sense.[10] What the cooperative miners show is how subterranean wealth moves through and undergirds multiscalar *patronaje* networks that extend from the nation-state to the individual cooperative. While property has always been the ultimate source of *patrones'* wealth and political power, in this case it is the dynamic between two different kinds of subterranean property that matters: natural gas fields (which are, for cooperative miners, abstracted hydrocarbon dollars) and mining concessions (concrete mineralogical formations to which cooperative miners want access). As in other contexts, patrimony is used to "assert claims over resources based on particular notions of the social group and its legitimate boundaries" (Ferry 2002, 333), and the expectations of national patrimony (immediate money) is different from the expectations of local patrimony (an enduring potential for wealth).[11]

The colonial dynamics of *patronaje* were particularly pernicious in the mines, eventually reinscribing themselves in the twentieth-century company town. For instance, most of the mining towns in Bolivia maintained a company store, or *pulpería*, that supplied deeply discounted basic food products, such as meat, bread, and rice, to all company employees. To miners, the *pulpería* was not just useful but also symbolic social infrastructure that the company was morally obligated to maintain. At the same time, however, preserving the *pulpería* allowed the company to keep artificially low salaries and internally circulating wages, which maximized company profits. Moreover, because miners could run up tabs in the store against their wages, their entitlements quickly transitioned into debts. Finally, since only properly salaried miners had access to the *pulperías*, the stores tended to deepen social divisions between salaried and subsidiary miners, the latter of whom paid for the right to exploit a particular section of the mine and sold what they could extract back to the company (see chapter 2). Similar stories of entitlement, debt, and social partition could be constructed in relation to company-provided housing, health care, and leisure venues such as sports arenas and swimming pools. In all cases, supposed gifts provided by apparently indulgent *patrones* established expectations of loyalty and repayment.

Even after the state-owned mining company COMIBOL closed its doors in 1985, this dynamic continued. Only the primary *patrón* had changed: miners now engaged directly with the state proper, without the company to act as a mediating institution. The declaration that announced the formation of the cooperatives in 1987 also announced that everything that had belonged to COMIBOL—including not only machines but also buildings, furniture, and structures internal to the mountain (elevators, trolleys)—would be gifted to the cooperatives. Cooperative miners would have full control over the equipment, and full responsibilities for maintenance and repair. But these gifts came with a hefty price tag. The newly minted cooperatives were expected to gradually repay the state for the machinery, with interest. Today, with every sale of mineral that the cooperatives make to commercializers, 0.5 percent is deducted as a payment on this old debt.[12] Given that the interest is greater than the cooperatives' monthly contributions, the debt only grows. The vice president of the cooperative Siglo XX, the largest and most indebted group of miners in Norte Potosí, told me in 2017 that his cooperative owed US$3 million for machinery and equipment (Llallagua, May 25, 2017).

Evo's election deepened rather than transformed this dynamic. After successfully uniting segments of the popular and working classes that would not normally have seen eye to eye, Evo maintained his multisectoral support in ways that could easily be (and frequently were) characterized as patronage. For instance, in 2006, after cooperative miners supported his first successful presidential campaign, Evo appointed Walter Villarroel, a cooperative miner from the tin-mining town of Huanuni, as minister of mining and metallurgy—a first for the cooperative mining sector. Although Villarroel was soon dismissed from this post following a massive (sub)territorial conflict between cooperative and union miners in his hometown in which seventeen people lost their lives (Howard and Dangl 2006), Evo continued to pay his dues to the cooperative miners in the form of *obras* and other gifts made possible by natural gas rents. Indeed, although cooperative miners usually called the president "compañero Evo," signaling equality and allyship, I also heard them call him "compadre Evo," a phrasing meant to remind him of his paternal coresponsibility for raising the nation's children. Nevertheless, while the exchange of political support for natural gas rents was consensual, Evo's gifts—like those of colonial and industrial *patrones*—often had a coercive aftertaste. The gifts came with debts that acted like leashes on the cooperatives, available for yanking whenever the miners stepped out of line.

Bolivian economist Roberto Laserna ([2005] 2011) argues that rent seeking, defined as attitudes and behaviors oriented toward obtaining benefits from collectively held natural resources, is one of the most defining and problematic features of Bolivian politics. He goes on to suggest that the contemporary tendency for Bolivians to engage politically as entitlement-demanding collectives rather than rights-bearing individuals is an inheritance of the colonial era, in which the population was hierarchically divided and separately regulated. Although similar to my history in that it centers the construction of the state as a father figure responsible for redistributing wealth among its children, Laserna's argument focuses more on the distribution of money from the state to collective actors and less on how this money filters down to and across each of these collectives. Bolivia's patronage networks are multiscalar: rather than flowing directly between president and voters, gifts are mediated by powerful actors at multiple levels—a phenomenon that occurs in many postcolonial nation-states.[13] In Bolivia this mediation is performed by social collectives that maintain internal forms of patronage that mimic the disciplinary fathering of colonial *patronaje*. The party system runs through the mining cooperatives, as it also does through the unions, which meant that the MAS government was more likely to patronize a regional cooperative mining federation with a group of MAS-affiliated leaders. The entire time I was in Llallagua, all the leaders in all the individual cooperatives and the regional office—like Esteban and Alfonso—professed allegiance to the MAS, despite unofficial opinions whispered around the offices. The FERECOMINORPO directorate acted as *patrón* to the leaders of each cooperative, who acted as *patrones* to their cooperatives' *cabecillos* (work crew leaders), who acted as *patrones* to their fellow work crew members or (illegally contracted) employees. The dynamic that Laserna identifies thus echoed across multiple geographic scales and levels of leadership.

The organization of the multiscalar patronage network within mining cooperatives has been shaped by both the history of the mining sector and the physical characteristics of the resource itself. As described in chapters 2 and 3, when mining cooperatives formed in the aftermath of economic neoliberalization, they were composed not only of previously salaried miners but also of former subsidiary miners and newly arrived migrants from nearby ayllus. A social hierarchy that had been external to the unions was suddenly internal to the cooperatives. Ex-salaried miners occupied positions at the apex of this hierarchy, a position they held on to by divvying up the richest subterranean sections among themselves. Although they were still, ultimately, renting their mining concession from the state, in day-to-day operations ex-salaried miners claimed the position of landlords and *patrones*, directing and filtering the flow of resource rents down to less powerful cooperative miners and across the community more broadly. As in the early colonial era, the maintenance of this power was made possible by the biophysical characteristics of the resource: minerals appear in rigid underground formations that can be parceled like three-dimensional plots, and while mining in one of these plots might affect the structural integrity of the mountain, it will not decrease the availability of minerals in another plot—minerals do not flow on human timescales. Just as important, the subsoil's legal separation from the topsoil and official framing as the property of all Bolivians, held in trust by the state, allowed cooperative miners to gain access to the entire subterranean concession all at once. Contending with only one landlord (the state), cooperative miners have much more extensive subterranean access than they would have if they were contending with the numerous owners of the lands above. *Patronaje* and patronage are thus colonial dynamics that depend on subterranean properties—the physical and the legal—to bind state and society together.

Back in the sports arena in Cochabamba, the cooperative miners from Norte Potosí were also whispering about *patrones* in their midst, but for them this title could only be reasonably applied to cooperative miners from the regional federation of Central Potosí. Hierarchies also run across federations within FENCOMIN, and Central Potosí, as the oldest and largest regional federation, had historically dominated national-level leadership positions. Moreover, while there had long been generalized grumbling about this inequality, Central Potosí's reputation had been particularly tarnished by the ill-fated roadblock in Panduro.[14] Central Potosinos, whispered the miners around me, were the real *patrones*: they never performed manual labor, preferring to use the money from the mines to pursue professional degrees or other business ventures. The imprisoned former leader of FENCOMIN was from Central Potosí, and the Norte Potosinos believed that he had ordered the whole roadblock just to protect his district from being further pressured to conform to the labor laws.

Tensions therefore were running high by the end of the second day of the congress, when the new directorate was due to be announced. Although officially an election, leadership decisions were made largely behind closed doors. Every mining cooperative across the country had put forward one representative, and each regional federation had chosen one or two of these representatives to become candidates for the FENCOMIN directorate. The presidents of the regional federations had subsequently locked themselves in a private room for the duration of the national congress to fight over the posts to which their candidates would be appointed. When the announcement finally began, everyone leaned forward in the bleachers. The speaker began at the bottom of the directorate hierarchy—the surveillance committee, the representative of sports—and worked his way up, getting each regional federation to "vote" by roaring "Aprobado!" (Approved!) after he announced the pick.

The presidency went to a miner from the federation Sur Atocha, located in southern Potosí. If this was not exactly a surprise, miners from Central Potosí had strong feelings about it. The only candidate from Central Potosí who made it to the FENCOMIN directorate was the representative of women, who was rarely involved in any real decision-making—Central Potosí, in other words, had been punished by having its representation feminized. But these miners did not accept the overthrow of their dynasty so easily, and they started storming down the bleachers toward the stadium floor. As they descended, miners from the other regional federations hurled insults at them; in return, the Central Potosinos began throwing plastic bottles up into the bleachers. These were thrown back at them, accompanied by a glass bottle that shattered across the cement. Miners from Sur Atocha began to climb down into the ring, and I readied myself to dash for the door if a fight broke out. The Central Potosinos, however, seemed to realize suddenly how outnumbered they were and withdrew angrily, allowing the election to continue. Sitting next to me, one of the Norte Potosinos nodded his head in satisfaction that the Central Potosí reign was over: "Ya no queremos patrones. Por culpa de ellos somos maltratados, malvistos." (We don't want bosses anymore. It's their fault that we are poorly treated, poorly understood.)

The patronage networks through which natural gas rents move are multiscalar and have been shaped by both colonial histories and geological materialities. As seen with the Central Potosinos, the uneven, filtered distribution of subterranean (natural gas) wealth can strengthen local hierarchies and exacerbate regional tensions. But these gaseous gifts also strengthened the cooperative mining sector as a whole—up to a point. As I have suggested, these gifts always came backed by the threat of violence of both physical and economic varieties. What happened in Panduro was well outside the established rules of engagement between mining cooperatives and the MAS government, and the miners were punished with bullets, incarceration, and financial exclusion. Such coercion is the flip side of a political dynamic built around subterranean redistribution through stratified networks of fictive kinship.

GEOLOGICAL VIOLENCE: PROTEST AND PUNISHMENT

Two weeks after Vice Minister Illanes's death, and two weeks before the congress in Cochabamba, I met with Alfonso and the rest of the FERECOMINORPO directorate in the courtyard of their office in Llallagua. The men were all stony-faced, mostly sitting on a bench with their arms crossed on their chests. Only Alfonso was standing, leaning on a broom. When he spotted me, he immediately began recounting his experience of the protest. Although he had not actually been at the front lines at the time of the murder—he had been in La Paz, negotiating the end of the roadblock—he had nonetheless been arrested, beaten, and forced to walk barefoot by police officers, who had fractured his ribs with their fists. He lifted his shirt to show me the bandages that circled his rib cage. Juan, the secretary of the directorate, picked up the story where Alfonso left off by saying that he, having remained at the roadblock while Alfonso was in the city, had witnessed two miners getting shot in the face with police bullets. He went on to enumerate the miners from the district who had been imprisoned, counting them off on his fingers one by one. The others nodded as he spoke, looking tense. I glanced around the circle and asked after Pedro, the vice president of the federation, who could usually be found in his office and whose absence at this moment struck me as unusual. An uncomfortable silence followed, finally broken by Alfonso's low, staccato words: "He was there. At the protest. With Illanes. He's gone underground now." I swallowed hard and nodded.

Alfonso was not speaking metaphorically: the literal underground has always been where miners go to hide from authoritarian or disciplinary eyes. The maze of underground tunnels is far more complex than can be seen on any engineer's map, a fact that has become increasingly true since the creation of mining cooperatives. Miners have created networks of passageways that are known only to them and their closest friends or family members, passageways where desperate miners could conceivably hide indefinitely, if someone occasionally resupplied them with food and water. The darkness of the subterranean, the apparently irrational layout of the mine, and the abundance of small, enclosed, and ultimately very private spaces all contribute to making the mine an excellent place to buy time and wait out a wave of potential violence.[15] I had heard tales of miners hiding underground from the police or military for months at a time, letting everyone assume that they had run for the border, only to reappear after their crimes had been forgotten. If the highways are the miners' offensive arena, the mines are their defensive zone.

I asked Alfonso and the rest of the directorate if it would be helpful for me to write an opinion article for a newspaper that showcased their perspective on the roadblock: why they had chosen this action, what they had hoped to gain from it, and how the subsequent violence had affected them. Although urban media had almost exclusively covered Vice Minister Illanes's death, five cooperative miners had also lost their lives at the roadblock in Panduro.[16] One of these men had died from mishandled dynamite, but police bullets were responsible for the other four. No investigation was ever launched into these deaths, and I had been unable to find more than scraps of information written about them in newspapers. Moreover, very little had been written about the fifty-nine miners arrested by the police, only a handful of whom had been formally accused of participating in Illanes's murder.[17] Most miners had fled the scene by the time Illanes's body was discovered, and the police were left to detain stragglers and latecomers. As I later discovered by speaking with imprisoned miners in the San Pedro prison in La Paz in October 2016, many of them did not even know about the murder for days after their arrests. Sharing some of these details in the public sphere, I proposed, might help counterbalance dominant narratives about the mining cooperatives' brutality.

Alfonso and the rest of the directorate, however, stared at me silently when I suggested this. Finally, Alfonso laughed mirthlessly. "It's that we're afraid," he explained. "We're afraid of being thrown in jail, and we're afraid they won't release those of us who are already there." Alfonso went on to explain that he thought the whole situation had been a setup because no one would have naively sent in a politician alone, with no military backup, to negotiate with miners. By repeating media discourses that framed cooperative miners as animals whose rage could not be contained without a military presence, Alfonso redirected the responsibility. If we're such animals, he seemed to be suggesting, then the handlers should have taken more care.

What did the miners plan to do next, I asked Alfonso, if not publicize their own version of the story? He explained that before they could pursue any action, they would have to decide as a collective whether they would keep fighting Evo and the MAS or whether they would crawl back to the president "with their pants down," as Alfonso put it, linking political humiliation to sexual submission. Alfonso seemed personally crushed with anxiety over which course of action he should encourage. He did not see a real path forward for the cooperatives without Evo's support but was painfully aware that he risked having his constituency turn on him if he continued to back Evo. Even I had heard rumors from other miners that Alfonso had been "negotiating" in La Paz during the protest because he knew something bad was going to happen. Why else, people whispered, would he have been released so quickly from prison, when so many were still languishing there?

Yet freeing the prisoners and clearing his name were only two of the challenges that Alfonso had to contend with. In the days immediately following the roadblock, the state had complemented its physical punishment with forms of harsh legal retribution. On September 1 Evo announced five supreme decrees designed to limit the cooperatives' political and economic activities. First, the use of dynamite in protests was prohibited (DS 2888). Dynamite is and has long been the readily available weapon of choice for miners, and this was also the case in Panduro. To ensure the new rule would be followed, dynamite sales were suspended nationwide, effectively halting miners' economic activities. Second, a new set of procedures to monitor and regulate mining cooperatives was introduced (DS 2889). In addition to acquiring a host of new licenses, mining cooperatives were now required to produce years' worth of economic data that, in most cases, had simply not been recorded. Third and fourth, the state repossessed all concessions registered to mining cooperatives that were either inactive (DS 2890) or operated by cooperative-private alliances (DS 2891). The final decree reiterated the right to unionization of all third-party workers employed by mining cooperatives (DS 2892), an announcement that only served to reinforce the modifications already made to the cooperative law. In response to critiques that these decrees were not robust enough to supplant rules contained in the existing Law of Mining and Metallurgy (Law 535), by October all five decrees had been elevated to the status of law (Law 845).

The punishments that were meted out against the cooperative miners not only were violent (in both physical and economic senses) but also, in many ways, harnessed the qualities of the subsoil against the miners. The state, as the ultimate landlord of the subsoil, had absolute power to evict its renters, to restrict the profitability of renting (by limiting collaborations with private companies), and to create complicated procedural barriers to renting in the first place. Because lode deposits require dynamite, moreover, restricting access to the explosive not only disarmed miners, which was the law's explicit intent, but also paused their productive capacities. All of these were reminders that the state—as patron, as collective father, as broker of a shared inheritance—could withdraw as well as provide support. As a lawyer hired to represent FENCOMIN explained to a crowd of assembled miners at the congress in Cochabamba, these laws seemed designed to "asphyxiate the cooperative system." If enforced, they could be as damaging to the cooperative system as silica dust was to a miner's lungs.

For Alfonso, the firm rejection of private-cooperative partnerships contained within DS 2891 and Law 845 was the most politically damaging. His major accomplishment during his first year in office had been attracting the attention of two foreign private companies—one Brazilian and one Chinese—interested in partnering with the cooperatives of Llallagua to exploit the ancient slag heaps that the latter had inherited from the state. The right to form partnerships with private companies had already been the subject of a major controversy between mining co-operatives and the state in 2014, when it emerged as part of a struggle over the new mining law (Law 535), but it had resurfaced in the list of demands presented by miners at the Panduro roadblock in 2016. As detailed in chapter 1, in 2014 the Senate failed to ratify a draft of the new mining law because Article 151a stated that mining cooperatives had the right to form associations with private companies, which the senators found unconstitutional since cooperative miners were supposed to be not-for-profit entities. At that time, cooperative miners had torn through the streets, shutting down several major transportation arteries for days, but were ultimately unsuccessful in their bid to reinstate the original article. Nevertheless, leaders like Alfonso had held out hope that the president might overturn this decision, a hope that was definitively dashed with the supreme decrees of September 1. These decrees negated the advances Alfonso had been making with the private companies and made him significantly less popular with the miners. They also reminded cooperative miners, once more, who was ultimately in control of the subterranean.

The new legal framework had a similar impact on Esteban's ability to claim that he represented cooperative miners in parliament. At the congress in Cochabamba, I witnessed Esteban attempt to explain the decrees article by article from the perspective of a member of the legislative assembly. But before he could get through the first decree, a younger miner stood up and shouted that Esteban had no right to talk, given how useless he had been during the conflict: "Usted es del sistema y debería llegar aquí con una propuesta, pero no" (You are from the system and should have come here with a proposal, but you didn't). Another miner yelled that Esteban was not even wearing his helmet, an important symbol that was technically required at all cooperative mining functions. His bare head showed how far removed he was from cooperative concerns. The floodgates had opened. Everyone began yelling at Esteban, who scrambled to find his helmet and protested that he was one of only a few people representing mining cooperatives in a parliament of 130—he could not exert *that* much influence on the lawmaking processes. No one wanted to hear his excuses, however, and he was forced to sit back down.

In the months after the Cochabamba congress, however, it seemed increasingly clear that cooperative miners were choosing the path of ingratiation rather than confrontation with the state: the new FENCOMIN directorate had settled into their offices in La Paz, public apologies had been issued, and no new protests had emerged—at least none large enough to appear in the national news. I was therefore surprised to learn that Alfonso's political position in Llallagua was no less fraught than it had been in early September. In January 2017 I encountered a small film crew standing outside the FERECOMINORPO office, interviewing Alfonso about a recent confrontation in the federation offices. Alfonso was describing the incident to the reporter as a "small problem" internal to the federation, but Juan, the directorate secretary, explained to me in a whisper that at least seventy miners had shown up, armed with dynamite, and had attempted to take the federation offices by force. They were upset, Juan added, because they thought Alfonso was taking advantage of his position as president of the federation to jump ship into the government. Alfonso had been promised a position in a national ministry—Juan didn't know which one—if he was able to pacify the miners in his federation who were still furious with the MAS government. After the interview Alfonso came into the office and flung himself in a red plastic chair, shaking his head and saying that the miners *de base* (of the rank and file) just didn't understand why he and the other regional federation leaders had to grovel to the government. He did not mention his own potential move into the ministry, and I did not ask.

I got the full story a few weeks later at the thirtieth-anniversary party of one of the smaller mining cooperatives, Carmen. Many cooperatives were celebrating thirtieth anniversaries in 2017, having been formed in 1987 in the immediate aftermath of neoliberal restructuring, and Alfonso was taking the opportunity to give a series of rousing speeches about the need for cooperative miners to make amends with the state, since only the state could keep them economically afloat. Over beers during the afterparty, he grinned and leaned in to whisper that by April his cooperative mining days would be over. Evo had personally promised him a vice-ministerial position in La Paz—although Alfonso, like Juan, didn't know which ministry. All he had to do was show Evo that Norte Potosí's cooperatives still supported the MAS. To that end, he added, he was heading to La Paz the following day to celebrate the Day of the Plurinational State (January 21). Seemingly on a whim, he invited me to join him, and, without dwelling on what the celebration would entail, I agreed.

I was not thrilled to discover that our van would leave at 4 a.m., fewer than four hours after Alfonso extended the invitation. I was so tired when I climbed aboard that it took me several minutes to realize that Alfonso and his son were the only two cooperative miners present. The other four travelers—two students, one schoolteacher, and one local politician—were some of Llallagua's most strident MAS supporters. Alfonso apparently needed to demonstrate his loyalty personally, since none of the other cooperative miners had been willing to join. I began to wish I had stayed in bed.

The ride to the city took seven hours rather than the usual four, since the van blew two tires so early in the morning that we were unable to find any mechanics to help us patch them, but the enthusiasm of the group never seemed to waver. We ended up flagging down a bus, arriving in El Alto instead of La Paz, and racing through the city to make it to the starting line of the parade before it was too late. We should not have been worried. Instead of walking directly into a moving parade, as we all expected, we were made to wait shoulder to shoulder in a seething mass of parade participants for four hours before we were allowed to begin marching. Packed so tightly that we could not turn around and speak to one another, we entertained ourselves by listening to Evo's speech on someone's cell phone radio. The parade would start when Evo had finished talking, but it seemed that the president was in no hurry. The sun slowly ascended over the tops of the tall buildings and began scorching us from above. At 3,700 meters above sea level, the midday sun is ruthless. Rivulets of sweat ran down everyone's cheeks, since no one could move their arms to shed layers, and still Evo kept speaking.

When it was finally time to march, Alfonso yelled at me to grab the other end of a banner that we unfurled as we went. The march, I was shocked to discover, was only one lap around the Plaza Murillo, where the presidential palace sits, and we were finished in less than ten minutes. Indeed, it felt like we were marching *for* the president, who stood on a raised platform in front of the palace, though of course there were also several cameras broadcasting our brief procession on national news. When we finished, Alfonso began excitedly telling the others that he had seen Evo pointing specifically at me, the one gringa in the parade, and I realized too late that Alfonso had invited me along as part of the spectacle, to make Llallagua's small contingent stand out as a devout municipality despite its meager numbers.[18] This devotion was nationalist in timbre but also found its object specifically in Evo. Although the march was technically to honor the day the constitution was approved in 2009, January 21 also happened to be the day that Evo was first sworn into office in 2006. There was an unavoidable conflation of a ritual celebration of the Plurinational State and a ritual celebration of the man, Evo Morales, the embodiment of plurinationalism.

And yet: participation in these events would not be without its material reward, at least from Alfonso's perspective. We did not spend seven hours in transit and four hours sardined into a parade corral just to celebrate Evo's ongoing reign. Alfonso made no effort to disguise his own interest in getting back into the president's good graces, both on behalf of FERECOMINORPO and for his personal career prospects. The patronage networks on which regional mining cooperatives relied would not return to operation without such a show of loyalty.

Alfonso, however, never received his just reward. He held on to his position in FERECOMINORPO for as long as possible, eventually relinquishing it in June 2017. When I saw him after that, he made no mention of the vice-ministerial position and instead invited me to follow him around from event to event in Llallagua and Uncía, where—to quote another miner—he was "still pretending to be a leader." If he had not happened to be the head of FERECOMINORPO during the crucial 2016 protest, it is probable that he would be a leader still. But Alfonso had been marked. Because he had been a leader during the roadblock, it seemed unlikely that Evo would invite him to occupy a position of power in La Paz, and since he had tried to reconcile with the MAS after the roadblock, it seemed unlikely that his fellow miners would elect him to another position of local power. He fell through the fault line of the mining cooperatives' strained relationship with the state.

MIXED BLESSINGS

The sports arena that had been under construction in Llallagua for as long as I had been working there officially opened for business on May 31, 2017. The arena had been framed to me on multiple occasions as a gift from Evo to FERECOMINORPO, but funding had abruptly dried up following the failed referendum of February 2016. Given the disaster at Panduro in August 2016, construction on the arena did not resume until April 2017, but then it seemed to proceed at breakneck speed. Almost before the basic structure of the building was complete, a local artist was painting the crests of each of the federation's sixteen cooperatives, miners were painting the bleachers yellow and white, and even I was entrusted to roll gray paint over the concrete floor. The cooperative miners needed my assistance, since it was crucial that the arena be ready for its inauguration in May: Evo himself was scheduled to do the ceremony in what would be his first appearance in Norte Potosí since well before the referendum. Given that a hallmark of Evo's leadership style was personal appearances in rural communities, his absence had been noted, and his reappearance promised to be the regional highlight of the month.

Evo was scheduled to arrive at two o'clock, and at noon the builders were still making sure the door handles were operational. Assuming the schedule would get pushed back, I made the mistake of stepping away to eat lunch; when I returned an hour later, the arena was completely packed. I joined a heaving crowd of people trying to push through the back door of the building but was soon immobilized with a sun hat shoved in my face, an elbow in my back, and hands bracing themselves on my shoulders. It took me so long to fight my way to the bleachers and find a comfortable perch that I missed most of the short speeches—by the mayor of Llallagua, the governor of Potosí, the plurinominal deputy of Norte Potosí (still Esteban), the new FERECOMINORPO president (no longer Alfonso), and the new FENCOMIN president (whose election I had witnessed in Cochabamba). I sat down in time to watch Evo take the microphone, but I still had to strain to hear what he was saying over the roar of the crowd.

Although the arena was billed as a gift for the mining cooperatives, Evo's comments dwelled almost exclusively on the precooperative era. He referenced the unions' involvement in the National Revolution of 1952, the massacre of San Juan in 1967 (the fiftieth anniversary of which we would be celebrating in a month's time), and the "white massacre" of layoffs in 1985, but he said not a word about the region's subsequent organizing into mining cooperatives. Moreover, he sported a miner's helmet but no *k'epirina*, the symbol of cooperative mining that I had seen him wear at other events. I interpreted these details to mean that the bonds between the mining cooperatives and the state, while healing, were still raw. The *cooperativistas* of Norte Potosí had not fully regained their patron's favor.

When he finished speaking, Evo turned and exited the side of the stage. He reappeared a moment later wearing professional-looking soccer gear, flanked by a team of tall men in matching uniforms. From the other side of the stadium, a team made up of local cooperative miners appeared, looking rather ragtag in comparison. Despite appearances, the teams were well matched and were tied 5–5 at halftime. Then disaster struck. Someone had walked around the perimeter of the stadium floor during the break and blessed it liberally with Coca-Cola, which caused

Evo to slip. The audience let out a collective gasp when he fell, followed by a sigh of relief when he stood back up. But the incident nevertheless marked the end of the game: Evo canceled the second half and departed without closing remarks. There were a lot of long faces among the miners as they left the building, and I found out later that Evo had promised a new semitruck to FERECOMINORPO if the miners succeeded in beating his team at soccer.

In this story, as in others, natural gas rents from eastern Bolivia were materialized in Llallagua as an infrastructural gift: the sports arena. Along the way, however, these rents also (re)materialized old social channels and dynamics, each of which have their own geological histories. Natural gas rents flow toward cooperative miners (and others) along patronage networks that reinforce local and regional hierarchies. These are dynamics that existed in previous formulations of the state. Although leaders of Indigenous communities, popular collectives, or workers' unions would not have been welcomed into the state in the colonial and republican eras (as they are now), they were still regularly "captured" by gifts of relative power and wealth—gifts they could keep only if they could prevent local discontentment from overflowing local borders. Violence marked and continues to mark the limit of benevolent redistribution of the subsoil.

Prior to entering (or attempting to enter) the institutions of the Plurinational State, Esteban and Alfonso were both formed through engagement with subterranean materials that constituted their livelihood and important sense of identity. Both labored within Siglo XX, the largest and wealthiest mining cooperative in Llallagua, both were the sons of unionized miners, and both identified primarily as mestizo, though both spoke Quechua and had familial ties to (somewhat distant) rural communities. In other words, they had been formed in ways that left them similarly positioned to enter the national political arena and act as conduits for the redistribution of geological rents back to Norte Potosí. Their divergent stories, however, reveal both the degree to which the structure of the Plurinational State has facilitated the participation of cooperative miners—among others—and the degree to which this participation is limited by much older dynamics of patronage and violence, themselves rooted in the construction of the subterranean as both collective and regional patrimony.

As Esteban's story in particular shows, being a local conduit for the redistribution of gas rents can result in political mobility, facilitating a leap into state and state-adjacent institutions ranging from parliament to ministries to state-owned companies. Even under favorable circumstances, it takes a lot of work to obtain and maintain these rewards, as shown by Esteban's continuous campaigning and, less successfully, by Alfonso's efforts to participate in the Day of the Plurinational State celebrations. Moreover, as Alfonso's story demonstrates, the rewards are never guaranteed, and much depends on an individual's ability to ensure local support in a context of aggravated resource regionalism. He was president of FERECOMINORPO at a moment in which cooperative miners were increasingly suspicious of Evo's ability to adequately perform his role as father and benevolent redistributor of natural resource wealth, and correspondingly invested in protecting their own resource base from taxation or state oversight. In this context, the loss of local support constricted the flow of gifts from the state, while the disrupted flow of gifts further reduced local support, a combination that ultimately catapulted Alfonso out of politics altogether.

The bonds that lash local leaders such as Esteban and Alfonso to the state were forged through discourses of collective patrilineal inheritance (patrimony), the transfer of material wealth (patronage), and the threat of physical or economic violence. All these bonds are capable of nourishing or strangling local politics and individual careers. All of them, moreover, have been shaped by geologically informed histories. Geology is everywhere, even in a sports arena. Both the money and the social structures through which it flows are rooted deep in the earth, buried in the bedrock of the nation.

During my time in Bolivia, I amassed a small collection of rocks that were gifted to me by cooperative miners. Some are dark black with an oily sheen, some are gray with brownish mottling, and the most beautiful one has dozens of clear quartz cylinders emerging vertically from the rock's tin-studded base. In Llallagua-Uncía it is common to find such rocks adorning the tables and shelves of cooperative miners at home and in their offices. My friend Demetria keeps hers in a cabinet otherwise reserved for fine dishware, though she took a couple out to show me when I visited her. In their craggy cavities, I noticed a couple dots of *mistura* (confetti) still clinging (see figure A.1).

More than samples of economic potential, these rocks are personally meaningful. Lined up on my desk, my rocks remind me of the people who passed them along. Demetria regularly blesses her rocks with beer and *mistura* to ensure the ongoing productivity of the mine from which they came. These material objects, conglomerates of sedimentary and igneous rocks, cracked loose from their surroundings with pickaxes and dynamite, are valuable even though they will never be delivered to the market as commodities. Demetria extracted them herself from the vein she uncovered in her early days laboring underground, when she was particularly eager to prove that women did not cause ore veins to disappear. Tin from that vein allowed her to feed six children and support a hospitalized husband, and it later earned the respect of her male *compañeros*. Men joined her *cuadrilla* (work crew), but she remained in charge because she had discovered the vein and labored in it first. She learned to handle a drill and ascended an internal hierarchy that does not usually privilege women workers. These rocks made her who she is today, and she honors that relationship.

Demetria's rocks are powerful reminders that matter cannot be cracked loose from its discursive moorings. My goal in this book has been to show how geological matters are both inseparable from the meanings that con-

FIGURE A.1. Demetria's tin ore. Llallagua, July 23, 2014. Photo by author.

stitute them and powerfully influential in political arenas at multiple scales. To do this, I have drawn on but reworked a historical materialist method. My reformulated approach takes as fundamental the histories of the materials that constitute the extractive economy: the tools, the laborers, and the subterranean itself. These materials, and their constitutive meanings, shape economic processes from the inside out. At the micro level, as workers transform rocks into commodities, the calluses, injuries, and even smells that adhere to their bodies carry specific meanings that divide workers along lines of gender and race. This transformation takes place in and through volumetric space, both belowground and aboveground, inflected as much by histories of extraction as by biophysical distributions of subterranean minerals. At the macro level, the construction of the subterranean as a nationalist space has shaped not only Bolivia's national economy (which relies on subterranean resources) but also its political institutions and processes.

Historical materialism, I have suggested, offers necessary but often insufficient conceptual tools to understand these dynamics, given that words such as *resource*, *tin*, and *value* cannot stretch to accommodate the palimpsest of meanings preserved in Demetria's cupboard. What I have called *material history* is attentive to matters and histories that seem to be outside of the economic sphere and yet nevertheless constitute and influence it. The production of economic value—regardless of what kind of production, although my focus has been on the extractive labor of mining—is also constitutively the production of raced and gendered difference as well as collective belonging at a variety of spatial scales. These processes are interior to, rather than externally articulated with, economic processes that include technological transformation, livelihood migration, and labor itself. This is certainly not meant to imply that economic processes and spaces are the *only* ones that are generative of such differences and identifications but rather that economic processes and spaces cannot be fully understood without taking these "extra-economic" material histories into account.

Like the rocks, mining cooperatives are far more historically complicated than they appear. In Llallagua-Uncía mining cooperatives sometimes seem to retain the nationalist spirit that was fomented in the Chaco War of the 1930s and promulgated by unionized tin miners in the years before and after the 1952 National Revolution. This spirit is on display when cooperative miners justify their claims to the subsoil in the language of national inheritance, when they reminisce about the importance of the tin mines to the nation's modern history, and even when they demand the redistribution of natural gas rents from the lowlands. In other moments they express their demands in the languages of community or regional autonomy, a tendency that becomes apparent when cooperative miners celebrate their identity as *agro-mineros* (agricultural miners), when they seek leadership positions in their natal ayllus, and when they fight tooth and nail to avoid any form of taxation that would not be returned to their municipalities. Apparently contradictory political impulses—themselves remnants of multiple historical moments—coexist within the mining cooperatives and sometimes even within individual miners. Tin is more than a metal, and cooperatives are more than collectives of small-scale miners: both are historical matters all the way down.

I have argued that it is precisely because of this ambivalence that cooperative miners offer a unique window into the complicated dynamics of resource nationalism and plurinationalism in Bolivia; the material histories of cooperative miners offer insights into the material history of the nation. Since its transformation from a republic into a plurinational state in 2009, Bolivia has promised political autonomy and collective land rights to its multiple constitutive Indigenous nations. To the degree that they were kept, however, these promises turned out to be surface deep: the subterranean had been constructed as and remained national space, managed by the central state. Benefiting from the turn-of-the-century commodity boom, Bolivia's supposedly revolutionary President Evo Morales turned subterranean resources into rents that could be redistributed across the country, legitimizing the extractive sector and his own political power—a textbook case of neo-extractivism, per Eduardo Gudynas (2009). In the introduction I described cooperative miners as fault lines in that, through their daily transgressions of surface/subsurface and nationalist/Indigenous boundaries, they embodied the tensions between divergent national projects. Following these fault lines reveals insights beyond the cooperatives themselves, since they are also intimately entangled with miners' unions, Indigenous ayllus, and popular economic movements.

Pressure that builds up slowly along fault lines is often released in short but catastrophic geological events. Using admittedly mixed geological metaphors, the remainder of this afterword charts three recent political *eruptions* that, I suggest, can be traced back to tensions inherent in the coexistence of resource nationalism and plurinationalism. Cooperative miners figure prominently only in the last eruption, but their stories—as recounted throughout this book—lend explanatory weight to all three analyses.

ERUPTION I: RESURGENT RESOURCE REGIONALISM

Much as it began, Evo Morales's presidency ended explosively. In October 2019 Evo ran for a fourth presidential term despite having lost the February referendum that would have given him the moral authority to do so. The constitutional court had subsequently granted him permission to run, but Evo's apparent disregard for the will of the people left him vulnerable to suspicion. Although early polling results gave Evo a clear lead, it did not seem that he would win by the 10 percent margin necessary to avoid a runoff against his main contender, centrist candidate Carlos Mesa. At this point in the vote-counting process, a technical error triggered a pause in counting that lasted nearly a full day. When the system came back online, Evo hurtled past the 10 percent mark, but the timing was fishy enough for the opposition to demand a recount.

The protests that followed the announcement of the election outcome were far from unified.[1] For reasons I have described throughout this book, Evo had already lost a significant number of his erstwhile supporters long before his name appeared on the 2019 ballot. Indigenous leaders, environmentalists, queer and feminist activists, anarchists, unionists, and committed leftists had all spoken out against Evo's fourth-term election on the grounds that he had made too many concessions to agrobusinesses, oil and natural gas industries, and mining interests. Yet losing the support of these relatively small, committed left-wing groups had never seriously harmed Evo's electoral chances. More numerically damning, enthusiasm had also begun to wane within the popular sectors, including campesinos, urban merchants, informal workers, cocaleros (coca growers), and cooperative miners. Support from these groups depended on Evo's ability to share the financial benefits of the subterranean, but given the falling price of natural gas, Evo's redistributive capacities had been arrested. Cooperative miners, for instance, continued to declare their official support for Evo and his party, the MAS, but they also regularly discussed the possibility of backing other, potentially more pliable candidates. They were not alone in their ambivalence, which was reflected in the February referendum and was once again displayed in the October election results.

Yet in the immediate aftermath of the election, the group that organized the most aggressive protests was neither the committed left nor the popular sector. Instead, a right-wing coup was made possible by widespread ambivalence among Evo's traditional devotees. Concentrated in the lowland departments of Santa Cruz, Tarija, Beni, and Pando, traditional right-wing elites had successfully captured the lion's share of the natural gas wealth over the previous decade, but they nevertheless seized the questionable election results as an opportunity to reassert their control over Bolivia's political sphere. Under the leadership of the previously unknown Santa Cruz Civic Committee leader Luís Fernando Camacho, they unleashed violence of a variety that had not been witnessed in Bolivia since the 2003 Gas War. Taking advantage of the fact that the streets of La Paz were already full of moderate and left-wing protesters demanding an election recount, Camacho and his far-right gang showed up with Bibles, baseball bats, and a resignation letter for Evo to sign.

Camacho framed much of his anti-Evo rhetoric in religious and cultural terms, yet the economic aspects of his crusade were hard to ignore. Camacho (a lawyer who had made his career supporting wealthy businesspeople seeking offshore bank accounts) and his supporters were eager to ensure that a greater share of natural gas rents remained in their home department of Santa Cruz. In this goal Camacho found an unlikely ally in Marco Antonio Pumari, president of the Potosí Civic Committee. While many analysts have argued that Camacho's right-wing gang seized and redirected the energies of a broad-based social movement to force Evo out of office, fewer have analyzed Camacho's alliance with Pumari.[2] In the days prior to the election, Pumari had led a twelve-day hunger strike to protest Evo's government plan to develop Potosí's lithium reserves through a joint-venture project with a German company. Demand for lithium is projected to skyrocket in the coming years along with demand for renewable energy technologies that rely on lithium-ion batteries. Thanks to the salt flats of southern Potosí, Bolivia is one of the three countries—the other two are Argentina and Chile—that compose the "lithium triangle," which contains an estimated 55 percent of the world's serviceable lithium resources. Potosinos are familiar with the downsides of hosting mineral booms, having been at the epicenter of both the colonial silver economy (central Potosí) and the industrial tin economy (northern Potosí). Now that southern Potosí was being framed by lithium boosters as the holy grail of the green energy transition, Pumari and his followers demanded that the MAS government guarantee that lithium rents would benefit *all* Potosinos, not just those immediately proximate to the salt flats. This led Pumari and the Potosí Civic Committee to join forces with Camacho's coalition.

It is in this alliance—between Camacho and Pumari, between natural gas and lithium, between lowland subterranean resources and highland subterranean resources—that it becomes clear how an articulated relationship between nature and nation permitted not only the rise but also the fall of Evo Morales. The 2003 Gas War that preceded his first election centered around a nationalist demand that natural gas rents be redistributed among all Bolivians. Evo's 2019 unseating, by contrast, was at least partly triggered by regionalist demands that subterranean rents be safeguarded in their departments of origin. But both moments were structured by a complicated relationship to the subterranean matters that undergird the nation physically, economically, and politically. This is certainly not to say that what happened was a "lithium coup" (as has been suggested), nor am I convinced that dissatisfaction with Evo was simply the result of declining natural gas prices in a country that had grown accustomed to this revenue stream. Instead, Bolivia's long history of subterranean resource extraction, resource imperialism, and resource nationalism shaped how people reacted to the political economic conjuncture. In both the lowlands and the highlands, the subterranean had long been at the heart of tensions between regionalism and nationalism, as well between right-wing and left-wing politics. This implies that the meanings and matters of the subterranean, especially natural gas and lithium brine, played a major role in the controversial election results—both the numerical results, in which Evo polled lower than he historically had, and the protests that emerged in the election's aftermath. The contentious connection between subterranean nature and nation was established well before Evo's time in office and will likely continue long after his departure.

These nuances faded in the weeks after the coup, as the interim government that took power was indisputably aligned with Camacho's traditional lowland elites. This government was helmed by Jeanine Áñez, a conservative senator from the lowland department of Beni, whose only official responsibility was to organize free and fair elections as soon as possible. Nevertheless, to quell protests against her administration—the legitimacy of which was hotly contested, given that she had sworn herself into office in the absence of a full parliamentary quorum—Áñez imposed a curfew and deployed the military throughout the country. Between twenty and forty people were killed in the town of Sacaba, outside of Cochabamba, and near the Senkata gas plant in El Alto, which was ironically also one of the major sites of violence in 2003. The MAS supporters went deep into hiding, while Evo himself fled to Mexico. State-backed violence continued in smaller waves for months while Áñez solidified her position, and the presidential election was not held until November 2020. While this election ended Bolivia's brief reactionary period by installing a new MAS candidate, current president Luis Arce, the reverberations of Áñez's violent regime continue to echo in the present.

Although the end of Evo's presidency was explosive, this eruption took place along multiple fault lines that were not all Evo's fault. His administration inherited not only a state that had been economically dependent on subterranean resource extraction for five hundred years but also a nation whose sense of unity had been built on a shared subterranean patrimony. These economic and political histories have shaped the country physically, leaving it riddled with mines and gas wells, and socioculturally, contributing to the production of spatially uneven social difference. These histories remain embedded in the very matter of the deep earth, where they continue to contour contemporary political imaginaries. While the growing contradiction between Evo's environmental discourse and his extractivist economics certainly exacerbated tensions, the tensions themselves were shaped by histories that are much older than his administration.

ERUPTION 2: COMMUNITARIAN MINING

Attributed to French anarchist Pierre-Joseph Proudhon, the declaration that "property is theft" has been a longtime rallying cry for leftist movements of varying political stripes. Much more recently, Robert Nichols (2020) has pointed out that since the very possibility of theft is predicated on the prior existence of property as a state-backed ideology, the phrase ought to be flipped into "theft is property." This is certainly true in Bolivia, where the theft of minerals—that is, mining outside of state regulations—was made possible by the colonial construction of the subterranean as sovereign property. This construction depended in turn on an imaginary of *sub terra nullius*, or the notion that the underground was both independent of the surface and devoid of human activity (Melo Zurita 2020). As explored in chapter 1, Indigenous nations were discursively and materially circumscribed to a shallow band of earth, an arrangement that has been preserved well into the officially postcolonial era.

In recent years, however, several Indigenous communities and activists in Bolivia have called this vertical separation into question. In 2013, in the lead-up to the approval of the new mining law, the highland Indigenous federation CONAMAQ organized a national summit to discuss the possibility of receiving formal recognition for what they called "communitarian mining." They hoped that communitarian mining could become a fourth "productive actor" in the mining sector, alongside the three constitutionally recognized groups (the state mining corporation [COMIBOL], private companies, and mining cooperatives). Although unsuccessful in 2013, this demand was raised again, nearly a decade later, in the context of rising mineral prices. On May 18, 2022, Luis Arce's government released a supreme decree (DS 4721) that included an article calling for the development of "policies that support Indigenous participation in the benefits of the mining industry."[3] Although neither the mining law nor the constitution has been altered to welcome a fourth productive actor into the mining sector, this decree seems to suggest that such legal harmonization might be possible in the future. Indeed, interviews conducted in June 2022 indicated that what was a relatively small demand in 2013 has since become a potentially major political eruption.

In the interim between communitarian mining's initial proposal and its current rise to prominence, I had the opportunity to interview one of the people who had led the original charge. Tata Félix Becerra had been the leader of the highland Indigenous federation CONAMAQ from 2011 to 2013. When we spoke in October 2016, however, he was imprisoned in San Pedro, La Paz's minimum-security prison for men, after having been accused of using money intended for development projects in his home ayllu for personal use.[4] His case was part of the scandal around the government's Indigenous Fund that exploded at the beginning of 2016 (see chapter 6), and he was one of at least a dozen Indigenous leaders to have been imprisoned after it was publicly revealed that the vast majority of fund-financed projects had never materialized. For Indigenous nations and their allies, this scandal and its accompanying incarcerations constituted an enormous set of smoke and mirrors designed to discredit Indigenous leaders who no longer supported Evo's presidency.[5] Tata Félix's leadership period had coincided with CONAMAQ's move to distance itself from Evo following the 2011 conflict over TIPNIS, the Indigenous territory and national park through which a highway was slated to be built, and animosities only deepened after he organized the communitarian mining summit in 2013.[6] In short, communitarian mining was the reason for my visit, but it was likely also one of the reasons Tata Félix had been imprisoned in the first place, given that it was one of several ways he had fallen out with Evo's government.

"Look," Tata Felix said when I had finished telling him about my interest in communitarian mining:

> The struggle for communitarian mining was always about the struggle for autonomy, about getting control over the nonrenewable resources in our territories. We are still doing communitarian mining, even though it was never approved. We work within the *marco* [legal framework] of the mining cooperativas, but we are self-governing Indigenous communities. We practice our *usos y costumbres* [uses and customs] within the mining process. For example, while a regular *cooperativista* will get a concession and then hire other workers to do the labor, in my sector all community members have automatic affiliation. Everyone who is over eighteen and who is participating in community responsibilities is given a lot in the

collective concession. Everyone is automatically affiliated to both the Indigenous federation and the mining cooperative. (La Paz, October 19, 2016)[7]

In other words, what was happening in his ayllu in Cochabamba was legally considered cooperative mining: the community had registered and received permission to operate as a cooperative. The difference between what they were doing and what a typical cooperative did, Tata Félix assured me, was in the details of their more egalitarian distribution of access rights and financial benefits. He explained that everyone in the community was assigned a three-dimensional "plot" on reaching adulthood—it was free as long as the person in question was performing their communal responsibilities. The only expense was a fixed amount paid annually (100 bolivianos, or about US$12.50) to cover the rental cost of the mining contract and some collective meetings. Everyone paid the same amount whether they were actively mining or not; that way the mine supported the community. This worked particularly well since some people had more agricultural responsibilities than others.

Félix wove his fingers together to show the inextricability of the ayllu and the mine, reminding me of the defense of cooperative mining to which I had frequently been treated by miners in Llallagua. The latter argue that the advantages of cooperative mining are its family orientation and its temporal flexibility, which allow miners to move regularly between the mines and the fields (see chapter 3). Like plots of lands, cooperative miners often treat mineralized veins as family property, such that a youth who wants to mine inherits the vein that their parent (usually their father) located and prepared. The significant difference here is that while cooperative miners might speak of their inheritance in terms of intergenerational labor and Bolivian citizenship, Tata Félix emphasized community belonging, which one might reformulate as Quechua citizenship. How much one's parent or grandparent had labored was irrelevant, as was belonging to the (colonial) nation-state.

Indeed, the more he spoke, the more it seemed Tata Félix was making an argument about Indigenous economic sovereignty, though this was not the language he used. In his words:

We produced coca, wheat, potatoes—but these have low prices. Slowly we decided we needed to administer our nonrenewable resources. Indigenous communities are smart; we wanted to be owners [dueños]. We knew our rights to manage our territory. We knew the laws, knew about contracts with international companies. In 2013 we wanted to move forward. That's when we had the [communitarian mining] summit. . . .

All minerals belong to the state, it's the owner [*dueño*]. [The state] didn't pay attention to Indigenous communities, and in the constitutional assembly they just associated us with llamas, *chuño* [freeze-dried potatoes], potatoes—they assumed we would live from that. But that's not the case! We have rights to *all* our resources. In 2006 we were already proposing communitarian mining, but in reality, we don't care what it's called. What's important is the idea of Indigenous autonomy: we can administer exploitation. It's territory all the way down [*territorio hasta abajo*].

This political vision shreds any romantic notion that Indigenous peoples are necessarily opposed to nonrenewable resource extraction; in fact, resource extraction can facilitate the economic autonomy that is so often missing from political autonomy. The demand for the profitable, nonrenewable resources muddies the line between "subsistence" and "extractive" communities, confusing progressive researchers and activists, but the appropriation of subterranean space for the purposes of economic sovereignty is not only practical but potentially decolonial. This is about not just *tierra* (land) but rather *territorio*, and it is a three-dimensional project.

Tata Félix's vision presents a challenge not only to neoliberal capitalism but also to resource nationalism and its more redistributive vision of shared national inheritance. Both neoliberal capitalism and resource nationalism rely on the spatial imaginary of *sub terra nullius*, which treats land and subsoil as two separate legal entities and erases any cultural or local idiosyncrasies from the latter. By challenging this division, Tata Félix and CONAMAQ revealed the limits of official interpretations of plurinationalism: although Indigenous nations could lay claim to ancestral territories along the surface of nation, those rights were only soil deep.

ERUPTION 3: PROLIFERATING THIEVES

Several years after completing the bulk of the research for this book, I returned to Bolivia for follow-up interviews in June and November 2022. Across all my conversations, I was struck by what seemed like a proliferation of people described by interviewees as "thieves." The number of mining cooperatives had grown precipitously, but it was more than that. Even in places without formal cooperatives, people talked about a "co-operative mentality" that manifested as theft (*jukeo*) or collective take-overs (*avasallamientos*) of established mines. The number of people who fell into the thief category had grown to include not only cooperative miners but also the children of salaried miners, campesinos, and Indigenous communities.

The causes of this proliferation are multiple, but a few are worth mentioning. First, the slow but perhaps accelerating global move away from hydrocarbons—encouraged by high oil prices as much as alarming climate change projections—has favored the metal market, given the material needs of renewable energy production. The prices of Bolivia's traditional metals (tin, silver, zinc, and lead) have trended upward at the same time that natural gas production has fallen. Even more important, the price of gold has soared as central banks and investors around the world hedge against global financial instability. Second, the COVID-19 pandemic had a particularly adverse impact on Bolivians working in the informal or popular sectors, both because these sectors lack structured health benefits (insurance, medical leave, and so on) and because it was impossible to continue such jobs remotely when the country was under lockdown (Hummel et al. 2021; McNelly 2021). Some of these people have sought to supplement falling incomes through mining. Third, the number of jobs in the mining sector has not risen alongside mineral prices, since industrial mining operations are increasingly dependent on large-scale technology rather than human labor. Instead of seeking formal employment, therefore, many people have looked for alternative ways to benefit from the boom in mineral prices.

As a direct result of these dynamics, the number of mining cooperatives in Bolivia has exploded. Most of this growth was in the gold-mining sector, which is based primarily in the eastern lowlands and the northern reaches of the department of La Paz. Bolivian gold mining has received a lot of attention because small-scale gold-mining operations are considered more environmentally and socially damaging than their traditional counterparts: cooperative gold miners are notorious for using large amounts of mercury, contaminating rivers in their search for alluvial deposits, displacing lowland Indigenous communities, and forming illicit partnerships with wealthy foreigners to acquire extractive machinery and distribution networks (Poveda Ávila 2021). The total number of mining cooperatives in the traditional sector has not increased as spectacularly, but membership has been creeping up within each coopera- tive. For instance, a friend from Llallagua estimated that the number of people working in the Juan del Valle Mountain had risen from 2,600 to 3,000 between 2020 and 2022. Moreover, the influx of people into gold mining has strengthened the cooperative sector as a whole, including its more traditional actors. Already a historically layered institution— an amalgamation of colonial-era popular economic strategies (*k'ajcheo*), early twentieth-century self-help strategies, Cold War union-busting tactics, and neoliberal restructuring—cooperative mining has now be- come an even more influential, widespread, and divisive sector.

The conjunctural contradiction of a growing number of people seek- ing to benefit from a sector with a decreasing number of jobs has also af- fected state and private mines, where salaried miners are outnumbered and, in some cases, geographically surrounded by cooperative miners. In the state-owned tin mine Huanuni, *jukeo* has also become a huge prob- lem, and most interviewees suggested that the culprits included both cooperative miners from Llallagua who washed the ore through their co- operatives and the adult children of salaried miners who were unable to acquire formal employment in the company (see also Arze Vargas 2021). In Colquiri, the other major state-owned mine, as well as in Bolívar, which is currently a private operation, salaried workers described feel- ing threatened by the growth of mining cooperatives, with which they share their deposits. The salaried workers fear *avasallamiento*, or seizure/ occupation, by *cooperativistas*—which is not uncommon, particularly in places where the number of cooperative miners is equal to or greater than the number of salaried workers. Animosity between salaried and cooperative miners has grown in tandem with the total number of co- operative miners, and analysts expect more conflicts in coming years. Cooperative miners, for their part, fear *avasallamiento* from *originarios* (original or Indigenous people) seeking to control extraction in their an- cestral territories. Rather than rejecting mining, an increasing number of Indigenous and campesino communities have found it more effective to seize the mine and control extraction themselves. A 2013 law against *avasallamiento* (Law 477) criminalized all mine occupations, regardless of their ultimate intent, but this has not stopped *avasalladores* (occupi- ers) in the context of rising metal prices. Instead, *avasallamiento* was a keyword in my 2022 interviews much more than it had been in 2016–17, suggesting a generalized fear of occupation among all actors within the mining sector.

Cooperative miners, *jukus*, and *avasalladores* all operate either on the margins of or fully outside the state's regulations. Yet they have found political strength in numbers: while mining companies and extractivist states certainly stand to benefit most from rising metal prices, a motley assortment of "thieves" is also figuring out ways to divert new financial flows. Cooperative mining, meanwhile, has gone from a marginal economic activity to one of the most powerful—albeit internally stratified—segments of society. Always an explosive sector, its newfound size suggests that future eruptions will be even larger than those of the past. The state's grip on the subsoil is loosening, and amid the variety of alternative claims, mining cooperatives constitute the most significant force.

How to think about the rise of small-scale mining, whether practiced by individuals, communities, or cooperatives, is relevant far beyond Bolivia. Whether impelled by factors such as a changing climate, environmental contamination, and land-use change, or attracted by the opportunities promised by rising metal prices, more and more people are incorporating small-scale and artisanal mining into their existing sets of livelihood practices. While radical theorists and activists have traditionally preferred to align themselves with wage laborers or smallholding farmers, this position is becoming increasingly untenable as global economic and environmental circumstances blur the boundaries among waged, self-provisioning, and informal economic practices. In this context, rather than politically unpredictable criminals, small-scale and cooperative miners might be imagined as presenting a wealth of political possibilities, capable of articulating with a range of other social and economic actors in the name of radical as well as reactive political change. There are no guarantees, but the challenges that such transgressive figures present to the sedimented-but-never-settled history of colonial and national resource extraction might be more productively explored than dismissed.

INTRODUCTION

1. This video was posted online by *El Nuevo Herald*, a Spanish-language newspaper based in southern Florida. "Aparece video que muestra a viceministro boliviano siendo amenazado de muerte" (A video appears that shows a Bolivian vice minister being threatened with death), *El Nuevo Herald*, August 30, 2016, https://www.elnuevoherald.com/noticias/mundo/america-latina/article98763332.html. The translation is mine, as are all others in this book unless otherwise noted.

2. The demand to partner with transnational mining companies reiterated a plea cooperative miners had been making since at least 2014 (Marston and Kennemore 2019). As Francescone (2015) details, this demand emerged as a result of state abandonment in traditional mining regions, which has left cooperative miners with few investment options.

3. On small-scale mining in Ghana, see Coyle Rosen (2020) and Luning and Pijpers (2017); on Brazil, see Cleary (1990) and de Theije (2020). Although cooperatives of small-scale miners have emerged in a few other countries in recent years—usually as part of formalization efforts, as in Peru and Brazil—Bolivia is unique in that the mining cooperative is the dominant form of organization for small-scale miners. Particularly given the importance of cooperative economics to leftist debates since Alexander Chayanov's ([1927] 1991) study of Russian peasant cooperatives, I suspect that my interlocutors were anxious to disabuse me of any romantic expectations I might have been harboring.

4. These numbers are notoriously hard to gauge, given that cooperatives grow and shrink in sync with mineral prices. The recent increase in prices since 2020 has caused a corresponding spike in the number of cooperative miners, with estimates now sitting closer to 200,000.

5. Although the precise number of independent, small-scale, and artisanal miners is difficult to track, one useful approximation is 40 million miners worldwide (Intergovernmental Forum on Mining, Minerals and Sustainable Development 2017), producing between 15 and 20 percent of global mineral output (Verbrugge and Besmanos 2016)—numbers that do not include secondary employment, dependent family members, or even quarrying of low-value subsoil resources such as gravel, sand, and limestone (Lahiri-Dutt 2018). Almost all tin (roughly 97 percent) comes from emerging and developing countries, especially

China, the Democratic Republic of the Congo, Brazil, and Bolivia, and about half of that is produced by artisanal and small-scale mining operations (International Tin Association 2020).

6. These two lefts are akin to Thea Riofrancos's (2020) two "resource radicalisms," which include "radical resource nationalism" and "anti-extractivism." However, I am less interested in periodizing the ascendancy of each "resource radicalism" and more interested in how the two tendencies coexist in imperfect harmony, rarely coherently distinguishable from one another within a single organization's political platform.

7. *Intercultural* is the term commonly used for Quechua and Aymara people who have migrated from the highlands to the lowlands, where they often practice agriculture—and increasingly mining—on a much larger scale than they did on the altiplano. Afro-Bolivians are mostly descendants of enslaved people brought from Africa to work in the silver mines and smelter of Potosí starting in the early 1600s. Today there are approximately twenty-three thousand Afro-Bolivians (0.2 percent of the population) (Zambrana B. 2014).

8. Brent Kaup (2010) describes it as a "neoliberal nationalization": constrained by the legacies of neoliberalism, Evo's administration regained majority control over the hydrocarbon sector through negotiated buyout agreements rather than through expropriation (see also Kohl and Farthing 2012). Even to this limited degree, however, nationalizing gas was a strategic move. Gas not only provides the greatest revenue stream in the country but is also symbolically linked to senses of nation and nationalism (Gustafson 2020), a connection that was particularly evident in the 2003 Gas War that unseated former president Gonzalo Sánchez de Lozada (Perreault 2006).

9. The International Monetary Fund's "Country Report" applauded the Morales government's "prudent fiscal policy," as Bolivia was one of the few countries in Latin America to maintain economic growth during the crisis of 2008 (International Monetary Fund 2014, 4). Bolivia still ranks low on the United Nations' Human Development Index, but the rate of extreme poverty has been dramatically reduced, largely through cash transfer programs. These also contributed to a near tripling of per capita income and overall improvement in health indicators.

10. See also elaborations on Latin American extractivism by Maristella Svampa (2019) and Fernanda Wanderley (2017).

11. See discussions of the TIPNIS conflict in Burman (2014), Fabricant and Postero (2015), Laing (2015), McNeish (2013), and Webber (2014).

12. On the ironic strengthening of lowland elites under Evo, see P. Anthias (2018), Fabricant (2012a), Gustafson (2020), and Postero (2017).

13. Lorismo refers to the work of Guillermo Lora, Bolivia's famous Trotskyist historian and political theorist, who was born and raised in the tin-mining region of northern Potosí. Leon Trotsky and Lora are popular among Bolivian miners because they argued that "semicolonial" countries did not need to pass through the same stages of development as "advanced" countries, since the semi-

colonized bourgeoisie was too compromised to undertake a proper bourgeois revolution (Ferreira 2010; John 2009).

14. This is the kind of historical or dialectical materialism that one would glean from reading *The Communist Manifesto* (Marx and Engels [1848] 1998) and never returning to read *Capital* (Marx [1867] 1990), let alone Karl Marx's many other treatises.

15. Many of the most meditative reflections on Marx's method have derived from readings of his 1857 introduction to the *Grundrisse* (Marx [1973] 1993). For some of these, see Hall (2003), Hartsock and Smith (1979), Ollman (2003), and Postone (1979).

16. Marx wrote, "The mode of production in material life determines the general character of the social, political and spiritual processes of life. *It is not the consciousness of men that determines their existence, but, on the contrary, their social existence determines their consciousness.* At a certain stage of their development, the material forces of production in society come into conflict with the existing relations of production or—what is but a legal expression of the same thing—with the property relations within which they had been at work before. From forms of development of the forces of production these relations turn into their fetters. Then comes the period of social revolution" ([1859] 1904, 11–12, emphasis added). In other words, the material forces and relations of production are what create "conscious" workers, who in turn become revolutionary subjects. The experience of labor conditions political actions and social outcomes.

17. Some of these critical new materialist accounts include Agard-Jones (2013), Alaimo and Hekman (2008), Barad (2007), Chen (2012), Z. Jackson (2020), and Murphy (2012).

18. I am indebted to the emerging field of "inhuman geography" (Clark 2011; Clark and Yusoff 2017; Yusoff 2017, 2018) and the broader rise of feminist geophilosophy (Bosworth 2017; Grosz 2008; Povinelli 2016).

19. This point has been made by Judith Butler (1993) and further developed by feminist and queer theorists, among others.

20. For instance, see Harvey (1982), Lefebvre ([1974] 1991), and N. Smith (1984). On Henri Lefebvre, see also Brenner and Elden (2009) and Merrifield (2013).

21. Eyal Weizman (2007) demonstrates how the Israeli occupation of Palestine has played out in aerial space; Peter Adey (2010), Stephen Graham (2016), and Francisco Klauser (2021) further develop this theme. Franck Billé (2019) examines aerial space in the formation of sovereignty, Jesse Rodenbiker (2019) demonstrates how the valuation of vertical space in Chinese cities contributes to social differentiation, and Julie Michelle Klinger (2018) looks at mineral speculation on the moon.

22. See Steinberg and Peters (2015) on thinking volumetrically about the ocean, Adler (2020) on the vertical dimensions of marine sovereignty, Woon and Zhang (2021) on suboceanic tunnel construction between China and Taiwan, Starosielski (2015) on undersea cable networks, and Jue (2020) on the ocean as a media environment.

23. Subterranean-focused works in political ecological and allied traditions include Bebbington and Bury (2013); Bobbette and Donovan (2019); Braun (2000); Bridge (2013); Kinchy, Phadke, and Smith (2018); Mendez, Prieto, and Godoy (2020); and Valdivia (2015).

24. Political geographic explorations of the subterranean include Billé (2020), Campbell (2019), Elden (2013), Hawkins (2020), Himley (2021), Klinke (2021), Libassi (2022), Marston (2019), Marston and Himley (2021), Scott (2008), Sorrensen (2014), Squire and Dodds (2020), Wang (2021), and Woon and Dodds (2021).

25. On the subterranean and the Anthropocene, see Clark and Yusoff (2017); Gerlofs (2021); Melo Zurita, Munro, and Houston (2018); Parikka (2015); and Yusoff (2018).

26. Works on the intimate or meaningful aspects of the subterranean include Ballestero (2019), Bosworth (2017), Marston (2021), Melo Zurita (2019), Oguz (2021), and Pérez and Melo Zurita (2020).

27. See also Rivera Cusicanqui (2018) and Van Buren (1996) on the possibilities and limits of this concept.

28. The most notable technologies were *huayrachinas*, natural draft furnaces, and *tocochimbos*, domed bellows furnaces (Van Buren and Mills 2005). Jeremy Ravi Mumford (2012) shows how the Spanish strove to govern in a way that preserved the "vertical" nature of the Incan Empire, which the Spaniards believed was a necessary feature of government at such precipitous heights.

29. See Galeano (1973), Kohl and Farthing (2006), Postero (2007), and Silverblatt (1987) for detailed accounts of the brutality of Bolivia's early colonial history.

30. The late 1700s have been characterized as a Latin American "age of insurgency" (Thomson 2002) or "age of insurrection" (Stern 1987) led by Afro-Latinx and Indigenous forces.

31. On race and *mestizaje* in Latin America, see de la Cadena (2000), R. Graham (1990), Sanjinés C. (2004), and Wade (1997).

32. For further elaborations on race and Lamarckianism in Latin America, see Marchesi (2014) and Nelson (2003).

33. Throughout this book, I capitalize "Indigenous" because it refers to an identity, but I do not capitalize "indigeneity" because it describes a sociospatial paradigm rather than a particular person or group. See Gustafson (2020) and Rifkin (2019) for similar discussions.

34. See Gustafson (2020) for a summary of the debates around the origins of the Chaco War. Although the rumor of oil proved largely inaccurate, the Chaco did end up containing a huge natural gas deposit that has become enormously important in recent decades.

35. See Rodríguez García (2012) and Lehm A. and Rivera Cusicanqui (2005) on the anarcho-syndicalists. Robert Smale (2010) traces the exchange between these anarcho-syndicalists and nascent miners' unions.

36. This is clear in Carlos Montenegro's *Nacionalismo y coloniaje* (Nationalism and colonialism, [1944] 1984), which argued that all of Bolivia's history could be narrated as a history of "nationalists" against "antinationalists." Montenegro

became a primary ideologue for the MNR, and his book exemplifies the nationalist tendency to claim previous Indigenous struggles as part of a nationalist history while obscuring ongoing Indigenous demands. See chapter 1 for further discussion.

37. On the historical rise of *katarismo* and the role of THOA, see Le Gouill (2014) and Rivera Cusicanqui (1987, 1992).

38. Ethnic-racial questions were avoided in censuses conducted after the 1952 National Revolution because all Bolivians had officially been declared mestizo in the postrevolutionary moment. This omission makes it difficult to compare the 2001 census results to earlier senses of identity. See Nicolas and Condori (2014) for a fascinating discussion on censuses in Bolivian nation-building projects.

39. In addition to the COMIBOL Archives (in Catavi and El Alto) and the National Archives of Bolivia (in Sucre), two of the richest archives I worked in belonged to Hans Möeller, a German-Bolivian economist who worked with FENCOMIN (Federación Nacional de Cooperativas Mineras de Bolivia; National Federation of Mining Cooperatives of Bolivia) in its early days. I also benefited from libraries held in the College of Geologists and the Museo Nacional de Etnografía y Folklore (National Museum of Ethnography and Folklore), both in La Paz.

40. These names are pseudonyms. Throughout this book I use pseudonyms for all cooperative miners and most other individuals, except those whose public activities make them readily identifiable.

CHAPTER 1

1. Some of the best examples of studies of legal and juridical contestation in Bolivia include Ellison (2018), Gustafson (2009b), Kennemore (2020), Postero (2007), Van Cott (2000), and Yashar (2005).

2. Law 535, art. 2.I, May 28, 2014, https://www.autoridadminera.gob.bo/public/uploads/Ley_535.pdf.

3. Technically, a mining company or cooperative must ask permission from its upstairs neighbor before beginning extraction, but this requirement is toothless. The 2009 constitution guarantees only "free, prior, and informed *consultation*" (CPE 2009, art. 352, emphasis added), not "free, prior, and informed *consent*," which is the wording used in both the International Labour Organization's Indigenous and Tribal Peoples Convention (C169, adopted in 1989) and the UN Declaration on the Rights of Indigenous Peoples (adopted in 2007).

4. In his historical analysis of Bolivian "beliefs" about natural resources, Fernando Molina (2011, 17) argues that three ideological beliefs have been particularly influential: natural resources in exchange for progress, natural resources in exchange for economic independence, and natural resources as a curse. He explores how these beliefs have been layered atop one another and interact with a more general "geological patriotism," defined as the tendency to celebrate the "quantity of mineral or petroleum wealth of the country, as well as its role in the world economy" (28).

5. d'Orbigny's first name is sometimes written as "Alcide" and sometimes as "Alcides," with the former more common in French-language texts and the latter more common in Spanish-language texts. I have adopted "Alcide" outside of direct quotations.

6. The question of whether and how Spain's imperial advances could be ethically undertaken had been under discussion since at least 1511, when Antonio de Montesinos, a Dominican missionary in Hispaniola, preached against abuses in the Americas. The Laws of Burgos, established in 1512, attempted to respond to these allegations by forbidding the mistreatment of Indigenous peoples in the Caribbean, but these rules were largely ignored in practice. In 1524 King Charles V of Spain established the Council of the Indies, which assisted in the passage of the New Laws of the Indies in 1542–43 and very nearly abolished the encomienda system. All these matters came to a head with the Valladolid Debate of 1550, after which the encomienda system was formally ended (Masters 2018).

7. See Anderson (2007) and Goldberg (2002) on intertwined histories of race and nature, particularly as pertaining to the Las Casas–Sepúlveda debate.

8. Mary Louise Pratt (1992) has shown how travel was a newly integral practice for European Enlightenment scientists of this time, whose practical challenge was that colonial powers did not often grant foreign scientists access to "their" colonies. The revolutionary wars that swept across Latin America in the first two decades of the nineteenth century changed this, as the leaders of newly inaugurated Latin American countries were eager to establish themselves as open for both economic and intellectual exchange.

9. God never disappeared altogether, but the power of his will was transferred to an orderly and rational natural world (Williams 1980).

10. That d'Orbigny was a French naturalist was also important, since postrevolutionary France was emblematic of secular, democratic modernity, and Santa Cruz had already been using France as a model for politics and knowledge production. The same year that he contacted d'Orbigny, Santa Cruz also issued a new Civil Code based on the Napoleonic Code, appointed three French men to the tribunal responsible for training health professionals, and made French language training mandatory in a series of new scientific high schools (Mendoza L. [1971] 2002).

11. Letters from June 10, 1830, and June 19, 1830. This correspondence is available online (d'Orbigny and Santa Cruz [1830] 2017).

12. See Pratt (1992) on European nineteenth-century explorers' dependence on the knowledge and labor of Indigenous guides in Latin America and Africa.

13. Although they appeared visually innovative, these vertical projections of mineralogical information were actually adaptations of "mining sections," a practical form of representing rocks that had been in use since at least the sixteenth century. Indeed, geognosy emerged at the Freiberg School of Mines under the guidance of Abraham Gottlob Werner, who was an influential figure for both Cuvier and Humboldt: Werner's geognostic studies inspired Cuvier's methodological approach to fossils, and Werner was Humboldt's mentor in the years prior to the

latter's expedition to the Americas. I have drawn most of my information about Cuvier's life from Rudwick (1997).

14. In this latter field, Cuvier had been highly critical of the Comte de Buffon, the former director of the Royal Gardens, for focusing only on externally visible characteristics; Cuvier believed that animals ought to be categorized on the basis of their *internal* organs. Cuvier developed sufficient dexterity with a scalpel to tease apart the internal organs of mollusks, which—as for d'Orbigny after him—were his first area of anatomical fascination. Fossils of mollusks, in turn, became Cuvier's key to dissecting geological strata and marking the passage of time (Rudwick 1997, 21n10).

15. Although Cuvier was careful not to contradict biblical genesis stories—after all, Napoleon had just declared himself emperor of France with the blessing of the pope—Cuvier's "geo-history" allowed for a great deal more elapsed time than had previously been imagined.

16. The epitome of Cuvier's contributions to scientific racism was the gruesome cataloging of Saartje Baartman's body parts—the last in a long line of cruelties suffered by this Khoikhoi woman from southern Africa. Baartman's scientific abusers were centrally concerned with how her features might help them answer the question of whether humans originated in one place and then migrated over the surface of the earth (as monogenesists claimed) or whether they emerged in multiple locations (as proposed by polygenesists). Cuvier belonged to the former camp, and he used Baartman's bodily measurements, along with copious other nonconsensual anatomical data, to "prove" that the "Ethiopian race" was fixed at the bottom of a racial hierarchy (Wallace-Sanders 2002).

17. This combination of empiricism and romance was so revolutionary that it was later hailed within anglophone literature as its own school of thought—a "Humboldtian" rather than "Baconian" science (Cannon 1978).

18. In the appendix of *El macizo boliviana*, where Mendoza detailed his sources, he notes, "Maybe the best study of the incipient field of Bolivian geology is also the oldest. It has been a century since a wise Frenchman, Alcide d'Orbigny, ascended the Bolivian Massif, and he studied it in large part by taking long trips, frequently on foot, for thousands of kilometers. . . . His work can be considered monumental" ([1933] 2016, 295).

19. On the relationship between gender and nationalism, see F. Anthias and Yuval-Davis (2005), Malkki (1992), and McClintock (1995).

20. See Coronil (1997) for a similar history of Venezuela.

21. DS 3196, art. 2, October 2, 1952, https://bolivia.infoleyes.com/norma/3058/decreto-supremo-3196.

22. While historians of Bolivia have traditionally portrayed the MNR as the unwilling servant of the US government, Kevin Young (2017) and Thomas Field (2014) have shown that members of the MNR were also frightened by the strength of the unions and stood to benefit from US intervention. This collaboration is particularly evident in the case of geology. In 1958 the Bolivian Ministry of the Economy struck a deal with the US Operation Mission to Bolivia (USOM)

that resulted in the formation of the Bolivian Department of Supervised Mining Credit, a program under which USOM granted loans to (private) Bolivian prospectors and developers, and in 1959 the program Desarrollo de Yacimientos Minerales (Development of Mineral Deposits) was created as the modification of a plan proposed by a firm of mining consultants engaged by USOM (Ford, Bacon, and Davis 1956). Financially supported by the United States, the program gave technical advice, loans, and even financial training to private mine operators (Kiilsgaard et al. 1992).

23. DS 7212, art. 3, June 11, 1965, https://www.vobolex.org/bolivia/decreto-ley-7212-del-11-junio-1965/.

24. See Law 535, art. 34, May 28, 2014, https://www.autoridadminera.gob.bo/public/uploads/Ley_535.pdf. Royalties are calculated in relation to the market price of the mineral in question. For tin, royalties are between 1 percent and 5 percent of the gross value of the sale (art. 227).

CHAPTER 2

1. There is a vibrant conversation that incorporates political ecologists, anthropologists, and new materialists about how to conceptualize the matter of resources, and what impact this matter has on people and politics. For helpful reviews, see Bakker and Bridge (2006, 2021) and Richardson and Weszkalnys (2014), as well as the introduction of this book.

2. According to David Harvey (2001, 25), one of the central contradictions of capital is "that it has to build a fixed space (or 'landscape') necessary for its own functioning at a certain point in its history only to have to destroy that space (and devalue much of the capital invested therein) at a later point in order to make way for a new 'spatial fix' (openings for fresh accumulation in new spaces and territories) at a later point in its history."

3. In a manner analogous to the anti–mining cooperative sentiment that prevails in Bolivia, international development and policy reports have traditionally framed small-scale mining as a problem requiring elimination. More recently, however, a growing body of ethnographic research has contributed much-needed nuance to this story. From a political perspective, James Smith (2022) demonstrates how small-scale coltan miners in postwar Democratic Republic of the Congo understand their livelihood practices to be contributing to peacebuilding, even and especially when their product is officially considered "bloody" due to its provenance in unmonitored former war zones. Economically, Daniel Tubb (2020) explores how small-scale gold mining in the Colombian Chocó has become one of several supportive livelihood practices for primarily rural Black Colombians. Numerous others, including Gavin Hilson (2016), Kuntala Lahiri-Dutt (2018), Matthew Libassi (2020), Nancy Lee Peluso (2017), and Robert Pijpers (2014), have made related arguments about the livelihood linkages that connect small-scale mining and smallholder agriculture. Environmentally, Pablo Jaramillo (2020) and Robert Pijpers and colleagues (2021) show how the category of "waste" is differently understood by small-scale miners, who endlessly reprocess tailings

and other mining "leftovers" that outsiders disparage (a topic to which I return in chapter 5). These studies, among others, have influenced more contemporary development efforts, which are now geared toward formalizing rather than eliminating small-scale mining—though this approach is also limited and has garnered its own critiques (Buss et al. 2019; Verbrugge and Besmanos 2016).

4. Argentine economist Raúl Prebisch, who led the famous Santiago-based Comisión Económica para América Latina y el Caribe (United Nations Economic Commission for Latin America and the Caribbean [CEPAL]), is perhaps most associated with the emergence of ISI.

5. Neoliberalism in Bolivia is a much-discussed topic, as the country was used as a testing ground for some of the most extreme structural adjustments. Although more sustained attention has been directed toward popular reactions to neoliberalism, see Kohl and Farthing (2006) and Perreault (2005) for detailed accounts of economic restructuring in Bolivia.

6. Numerous ethnographies have featured laid-off miners as important urban actors in Bolivia. See Gill (1997), Goldstein (2016), and Lazar (2008) for accounts of ex-miners participating in community organizations, local popular economies, and national politics. Many of these miners take pride in retaining their "formation as a worker" despite being structurally closer to the petty bourgeoisie or lumpen proletariat (Gill 2000, 80).

7. This should not be overstated: the *cocalero* unions that formed in the Chapare, while complicated in ways similar to the mining cooperatives, are far removed from cocaine cartels. Coca—like all raw resources—fetches little compared to the refined product into which it is sometimes transformed, and coca farmers are the poorest actors along this commodity chain (McSweeney et al. 2018). Moreover, their historical connection to miners' unions shows in their political actions. They often align themselves with workers' unions despite not being salaried workers themselves (Grisaffi 2018).

8. Gold mining in Bolivia does not follow the same historical arc associated with the traditional sector, which includes tin, silver, and *complejo* (mixtures of zinc, lead, and silver) mines. The differences between gold cooperatives and traditional cooperatives were so acute that the gold cooperatives split from FENCOMIN in 2015 and formed their own organization. Since the huge spike in gold prices that started in 2020, gold-mining cooperatives have become the largest subgroup of the cooperative sector in terms of membership and economic impact. They are also the most environmentally concerning, given that they work on rivers and use large quantities of mercury.

9. See also Taussig (1980) on Bolivian miners making deals with the devil.

10. Bolivian miners, unionized and cooperativized, are famous for paying tribute to a devil figure who controls underground spaces and is called El Tío (The uncle). See Absi (2005), June Nash ([1979] 2003), and Taussig (1980).

11. Edward Andrew (1983) notes that Marxists infer the difference between "in itself" and "for itself" from two passages written by Marx—in *The Poverty of Philosophy* ([1847] 2008) and *The Eighteenth Brumaire* ([1852] 1963)—neither of which

is explicit in its differentiation. For this reason, Marxists are divided on whether the distinction ought to exist at all, or what utility it serves (see also Neilson 2018).

12. Stuart Hall's (2003) reading of Marx's methodology insists that Marx's dialectical approach assumed a set of internal relations between economic theory and empirical history, such that what appears to be a claim of expressive causality must in fact be understood as an internally related process. See also Hart (2018) and Ollman (2003).

13. See John (2009) on the history of Trotskyism in Bolivia. Although Leon Trotsky is no longer as popular in Bolivia as he was in the mid-twentieth century, he remains a touchstone for trade unionists and scholars of mining specifically. Among my interlocutors in Llallagua-Uncía, Trotsky was referenced far more frequently than, for instance, the Bolivian Marxist economist René Zavaleta Mercado, despite the latter's comparative popularity among Bolivian and foreign scholars.

14. Centuries later, Bolivian writer Augusto Céspedes ([1946] 1986) popularized the nickname "devil's metal" in his fictionalized account of Patiño's life, but it is unclear if he was riffing on the Latin phrase or just referencing El Tío, the devilish figure worshipped in all subterranean mines in the Bolivian highlands.

15. This etymology was explained to me by the *palliris* themselves, but I also found it recorded in a publication by Bolivia's Vice Ministry of Women (Canaviri Sirpa and Selum Yaveta 2005, 8). While I am no linguist, it is curious to note the similarity between the word *palliri* and the French word *orpailleur* (gold panner; see d'Avignon 2018). *Orpailleur* is etymologically explained as a combination of *or* (gold) and *harpailler*, which is Old French for "to grab." The Latinized Quechua word *pallar* has also been translated for me as *agarrar*, or "to grab." I leave this colonial linguistic puzzle for someone else to decipher.

16. Edgar Ramírez of the COMIBOL Archives was especially committed to underscoring this point to me.

17. *Jukus'* demands in twentieth-century tin mines constituted a kind of moral economy (Thompson 1971) and resembled colonial-era practices of *k'ajcheo* in the silver mines of Potosí (Barragán Romano 2017).

18. There are several stunning historical accounts of the *k'ajchas*, or those who practiced *k'ajcheo*, but foremost among these are Abercrombie (1996), Stern (1988), and Tandeter (1992). See Absi (2006a) and Barragán Romano (2015, 2017) for specific connections between *k'ajchas* and cooperatives.

19. Nevertheless, the figure of the semilegal *k'ajcha* outmaneuvering wealthy Creole prospectors is a compelling symbolic ancestor to the contemporary cooperative miner, particularly in the iconic colonial city of Potosí (Abercombie 1996; Absi 2006a).

20. Godoy (1985b) includes *jukus* and *palliris* in his list of subsidiary organizations, but I have chosen to discuss these two groups separately both because they

were not properly "organizations" and because they have much longer histories than the other three types.

21. At the time, Juan del Valle was only one of several regional cooperatives, but the rest are now defunct, with apparently no trace that they ever existed. A former leader of FENCOMIN shared his copy of the *Recopilación Histórica de Congresos de la FENCOMIN* (Historic collection of the congresses of FENCOMIN [FENCOMIN 2008]), which was published to mark the federation's fortieth anniversary. This compendium showed that when FENCOMIN held its second national congress in 1970, there were ninety-eight mining cooperatives nationwide, nine of which were based in Norte Potosí. These nine were associated with the Oruro Departmental Federation, which is geographically closer than the city of Potosí. Norte Potosí's own regional federation, FERECOMINORPO, was not formed until 1987. Of these nine, one was Juan del Valle, but there is no record of the others. I repeatedly inquired about these cooperatives, but most people insisted that they had never existed. To me, the very ephemerality of these organizations points to how rural people used them to gain short-term or intermittent access to the subsoil (see chapter 3).

CHAPTER 3

1. Throughout the Andes, locally controlled radio stations were an important means of communication for miners and peasants organizing against oppression in the twentieth century (O'Connor 1990). In Llallagua the two prominent radio stations were La Voz del Minero, founded by unionized miners in 1952, and Radio Pio XII, a Catholic radio station run by Oblate missionaries from Canada. Radio Pio XII was initially at loggerheads with the unions, since the priests aimed to draw miners away from communism. But after they witnessed military brutality against the miners, and became inspired by the regional rise of liberation theology, the radio station began to champion workers' rights (López Vigil 1985). In the 1980s Pio XII also began working closely with the regional peasant unions and ayllus (Le Gouill 2014). Today it remains firmly on the side of "the people" but walks a fine line between supporting agricultural communities in their struggle against mining pollution and supporting mining cooperatives in their struggle for better livelihoods. Their daily radio program for miners is also called *K'epirina*.

2. In his classic book *Technics and Civilization* ([1934] 2010), American philosopher Lewis Mumford points out that mining is a strange occupation in that its contributions to modernity were essential, yet its associated tools (at the time he was writing) had changed little since premodern times.

3. Bolivian scholars Fausto Reinaga (1969) and Silvia Rivera Cusicanqui (1987, 2003) make this point, as do foreign scholars Anders Burman (2014), Andrew Canessa (2014), and Nancy Postero (2007, 2017), among others.

4. Historian Patrick Wolfe (2006) influentially argued that settler colonialism proceeded through a "logic of elimination." He contrasted the history of Native elimination in the United States to the concurrent history of Black exploitation,

in which the ongoing availability of cheap labor relied on the reproduction of a racially segmented society. The number of people in the United States legally categorized as Black (and subject to correspondingly exploitative legislation) grew continuously because of the infamous "one-drop rule," in which any Black ancestry resulted in Black identity, while in the same period, the number of people legally categorized as "Indian" decreased because the category was restricted to those who could identity a minimum number of Native ancestors. See Coulthard (2014), Simpson (2014), and TallBear (2013) on settler colonialism in Canada and the United States and Ybarra (2018) on settler colonialism in Guatemala.

5. Jo Guldi (2022) uses the phrase *indigenous labor colonies* to distinguish such places from settler colonies. She focuses on Ireland and India, but the same phrase could be used in Bolivia.

6. The doubled figures of the *indio permitido* (permitted Indian) and the *indio prohibido* (prohibited Indian) that Charles Hale (2004) describes emerging in the era of neoliberal multiculturalism might be traced back to the distinction between *indios* and *cholos*, or racial purity and racial mixture. Hale draws on Bolivian scholar Silvia Rivera Cusicanqui to make this distinction.

7. Ayllus are noncontiguous, spatially and politically nested polities with often-overlapping borders. This means that I received many different answers to the question of how many ayllus bordered the towns. While Chullpa, Kharacha, and Aymaya were consistently mentioned by my interlocutors, sometimes people mentioned other ayllus whose proximity would then be denied in subsequent conversations. I tried to use Fernando Mendoza Torrico and Félix Patzi González's *Atlas de los ayllus del Norte de Potosí, territorio de los antiguos Charka* (1997) to establish a fixed sense of the region's ayllus but was dissuaded from this project by Rivera Cusicanqui's (2015, 68) assertion that efforts to pin ayllus down geographically in this way are both masculinist and colonial. For the purposes of this chapter, therefore, I refer only to the three ayllus that everyone agreed were proximate and influential, and I have not attempted to map their relative distances from the towns.

8. Of the remaining twenty-five workers in my archival sample of three hundred employee files, nine were from Chuquisaca, one each was from Santa Cruz and Tarija, six were from foreign countries (Argentina, Chile, Peru, and Yugoslavia), and the rest were either illegible or blank. Results taken from the COMIBOL Archives in Catavi, a submunicipality of Llallagua.

9. Many miners made this point, though it was made particularly forcefully in an interview with Don Alfonso in Llallagua on February 25, 2016.

10. Unionized miners' wives took company-store meat and grains to rural markets to exchange for potatoes and vegetables (Godoy 1985b; Harris 1989).

11. Many thanks to Jen Rose Smith for encouraging me to pursue the potato and to Donald Moore for discussions about the postcolonial potato.

12. Drawing on fieldwork in the ayllu Laymi of Norte Potosí, Olivia Harris (1989, 253) notes that *llallawa*, the name of "strange-shaped tubers or maize cobs, stands for abundant harvest."

13. According to the International Potato Center, based in Lima, there are over four thousand edible varieties of potatoes in the world, and almost all are found in the Andes. International Potato Center, "Potato," accessed March 16, 2020, https://cipotato.org/potato/.

14. I was told that the best potato in Norte Potosí is the round *imilla*, identifiable by the dozen or so eyes that must be cut out with skilled knifework. Drawing on fieldwork conducted in Norte Potosí in the 1980s, Harris (1989) notes that *papa imilla* was a variety destined for the market, while other varieties were more often consumed at home.

15. The English word *potato* is a legacy of these Spanish sailors. Although in Bolivia the word for potato is *papa*, which is taken directly from Quechua, in Spain the word is *patata*, which comes from the Taíno word *batata*. Taíno is the predominant Indigenous language of the Caribbean, and *batata* refers to the Caribbean sweet potato (*Ipomoea batatas*), which the Spanish first ate when they landed in Hispaniola at the end of the fifteenth century. When they encountered savory potatoes (*Solanum tuberosum*) in the Andes some sixty years later, they applied the same word (Reader 2008; Salaman [1949] 1985).

16. What political economists often call the agrarian question, Bernstein (1996) argues, is more accurately a series of questions about the effects of capitalism on agriculture: an accumulation question (To what extent can agricultural surplus facilitate industrial expansion?), a production question (How does agriculture itself become capitalist?), and a political question (How does rural class composition shape political struggle?).

17. See Akram-Lodhi and Kay (2010a) for a review of historical Marxist thought on the agrarian question; and Akram-Lodhi and Kay (2010b) for an analysis of more contemporary debates.

18. The army physician had been imprisoned in Prussia, where he found his steady diet of potatoes surprisingly salubrious.

19. As part of a larger argument about the genealogical importance of subjugated knowledges, Michel Foucault ([1976] 2003) argues that class-based tensions evident in eighteenth-century France were actually an evolution of a seventeenth-century discursive "race struggle" that pitted the Celtic Gauls against the Germanic Franks, who had invaded in the first century CE.

20. The work of demonstrating the political consciousness of Latin American peasants has already been done by scholars such as Hugo Blanco (1972), Florencia Mallon (1994), and Steve Stern (1987), among others.

21. See Barragán Romano (2017) and Abercrombie (1996) for discussions of the racial and economic transgressions of miners in eighteenth-century Potosí.

22. The number of cooperative miners in the country is typically between 120,000 and 200,000, but Carlos was likely including dependent family members and/or indirect jobs (food service, ore transportation) in his estimate.

23. Personal conversations with antimining activists in Oruro, June 22, 2022, and La Paz, June 28, 2022.

24. This conflict was part of a major international scandal. Since 1999 the

Huanuni mine had been shared by four cooperatives and the US- and British-based company RBG / Allied Deals. In 2002 RBG / Allied Deals was charged with fraud in the United Kingdom after it was discovered that the company had been operating a global mining Ponzi scheme. On hearing this news, the salaried miners in Huanuni expelled the company from the mine, and Bolivian authorities began the process of dissolving the joint-venture contract. On the other side of the ocean, however, the British judicial system granted the consulting company Grant Thornton responsibility for liquidating RBG / Allied Deals' assets. Grant Thornton told Huanuni's mining cooperatives that they could have the concession if they could pay US$1.1 million, and the cooperatives purportedly agreed to this sum. This purchase was not considered legal by the Bolivian state, however, and the conflict that emerged between cooperative and salaried miners could only be resolved with a full nationalization of the mine. Altogether, more than four thousand cooperative miners and eight hundred ex-employees of RBG / Allied Deals became COMIBOL employees in Evo's government's first "renationalized" mine (Ruiz Arrieta 2012; interview with Héctor Córdova, La Paz, June 20, 2022). It remains the only operational mine to have been fully nationalized; a similar conflict in Colquiri in 2012 was resolved with a nationalization that incorporated only some of the mine's cooperative miners rather than all of them.

25. I take this phrase from Postero's (2017) book of the same title, but it has also been used by Canessa (2014) and Karl Zimmerer (2015), among others.

26. For example, Young (2017) explicitly attempts to construct a history of resource nationalism *outside* of the miners' unions since the association between mining and nationalism is already so well studied.

CHAPTER 4

1. Foundational Third World and intersectional feminist texts insisted that there was not a universal experience of womanhood and that being a woman of color and/or a woman in the Global South was not simply an experience akin to "more patriarchy," as white feminists believed (Crenshaw 1991; Mohanty 1988; Sandoval 1991; K.-Y. Taylor 2017). Some contemporary scholars have moved away from the language of intersectionality, which dominated in the 1990s and 2000s, to elevate Black and/or women-of-color feminisms (McKittrick 2006; Jennifer Nash 2019; K.-Y. Taylor 2017).

2. "If money, according to [Marie] Augier, 'comes into the world with a congenital blood-stain on one cheek,' capital comes dripping from head to toe, from every pore, with blood and dirt" (Marx [1867] 1990, 925–26).

3. I do not use the term *deformation* as the converse of *formation* because *deformation* has a particular meaning in geological sciences. It describes the change in the form or size of a rock caused by biophysical forces of heat, pressure, moisture, or gravity, none of which have any sort of negative connotation. *Degradation*, by contrast, is used by resource economists and environmentalists rather than geologists, and it describes a harmful process almost always caused by humans.

4. Debates among geographers in the 1980s and 1990s over whether nature is "produced" by society's capitalist relations (Castree 2000; N. Smith 1984) or "constructed" by society's more wide-ranging social relations (Demeritt 2002; Gifford 1996) sometimes missed the point that nature also produced/constructed society. Since the early 2000s, however, *coproduction* and *coconstitution* have become ubiquitous keywords among political ecologists and other geographers (Budds and Hinojosa 2012; Guthman 2011; Heynen et al. 2007; Lave et al. 2014; Swyngedouw 1999).

5. June Nash ([1979] 1993) reflects on her surprise at learning so much more about political economy while conducting fieldwork with miners than she had in her doctoral courses.

6. In Spanish the words *consciencia* (consciousness) and *conciencia* (conscience) are pronounced the same way, so when I read my field notes later that day, I was not certain which one the professor had meant. I found this linguistic indeterminacy useful: to be conscious is to have a conscience.

7. Gramsci was building on Marx's ([1852] 1963, 15) point that "men make their own history . . . but under circumstances directly encountered, given and transmitted from the past. The tradition of all the dead generations weighs like a nightmare on the brain of the living."

8. Although not all the following are self-identified political ecologists, they all explore how the matter of resources comes to matter in political and/or economic processes: Bakker (2003), Le Billon (2001), Mansfield (2003), Mitchell (2011), Robbins (2012), Sneddon (2007), and Watts (2001).

9. "Dirt is the by-product of a systemic ordering and classification of matter, in so far as ordering involves rejecting inappropriate elements" (Douglas [1966] 2003, 36).

10. See also Mary Weismantel's (2001) discussion of dirt in Peruvian marketplaces and Nancy Leys Stepan's (1991) analysis of mid-twentieth-century hygiene campaigns for racial "improvement" across Latin America.

11. Nash ([1979] 1993, 54–55) shares the story of one *palliri* who was similarly promoted to underground work during the Chaco War, only to be returned to the slag heaps at the war's end.

12. Similarly, Marisol de la Cadena (2000, 5) shows how educational achievements are used in Peru to explain the difference between indigeneity and *mestizaje* in supposedly nonracial terms.

13. Anthropologist Mary Weismantel (2001, 113) even writes about women in Peru who yearn to become *cholas* but cannot afford the proper clothes.

14. On gender and the military, see also Gill (2000) and Radcliffe and Westwood (1996).

15. Compare this to Povinelli's memorable comment on liberal multiculturalism: "In Western Europe and the United States, public anxieties about cultural diversity and national identity are often expressed at the tip of the clitoris" (1998, 575).

16. For comparison, see Stockwell (2015) on Canadian mining towns.

CHAPTER 5

1. "Owing to its conversion into an automaton, the instrument of labour confronts the worker during the labour process in the shape of capital, dead labour, which dominates and soaks up living labour-power" (Marx [1867] 1990, 548).

2. The plaque on the side of the building indicates that this renovation was funded during the presidency of Carlos Mesa, a liberal historian who served as vice president (2002–3) and interim president (2003–5). Apparently, Mesa took advantage of his brief time in the president's office to elevate the country's historical landmarks.

3. APEMIN stands for Apoyo al Desarrollo Económico Sostenible en Áreas Mineras Empobrecidas del Occidente de Bolivia (Support for Sustainable Economic Development in Impoverished Mining Areas in Western Bolivia), and EMPLEOMIN stands for Empleo en Áreas Mineros de Bolivia (Employment in Mining Areas of Bolivia).

CHAPTER 6

1. The details of this story unfolded slowly over the next year, and many elements of it remain unclear. At the time of writing, it seems that a child never existed. In 2017 Zapata was charged on several counts, including earning illicit income, maintaining a false identity, and misusing public goods, and is currently serving a ten-year prison sentence in La Paz (Carrasco 2017).

2. The duplicity of Latin America's bourgeoisie was the subject of much scholarship, particularly in the 1960s and 1970s. Marxist scholars argued that there could be no "bourgeois revolution" in Latin American countries because the bourgeoisie had earned their wealth by colluding with imperial powers rather than building their own capitalist businesses. Andre Gunder Frank (1972) called these elites the "lumpen bourgeoisie" because they failed to develop their own class consciousness.

3. Even Potosí, usually a MAS stronghold, voted "no" to the constitutional amendment. Norte Potosí voted "yes," but not with enough enthusiasm to outweigh the rest of the department (Órgano Electoral Plurinacional 2016).

4. This was the third time the natural gas sector was nationalized; the first was in 1937, on the heels of the Chaco War with Paraguay (1932–35), and the second was in 1969 (Gustafson 2020).

5. See P. Anthias (2018), Gustafson (2020), Hindery (2013), Humphreys Bebbington and Bebbington (2010), and McNeish (2013) on the complicated dynamics of "resource regionalism" in la media luna.

6. When it was first established in 1997, CONAMAQ was funded mostly by NGOs, which allowed the federation to express antigovernmental positions even while constraining it in other ways (Albro 2006; Andolina, Radcliffe, and Laurie 2005).

7. Another defining feature of the plurinational assembly is its emphasis on gender equality: 50 percent of the candidates on the plurinominal lists must be women, and any man running for an uninominal seat must run alongside a fe-

male alternate (and vice versa). As María Galindo (2017) argues, however, such nominal gender equality has not resulted in the introduction of policies benefiting women.

8. Of course, there is a difference between the goals of roadblocks in urban areas and those in rural areas. Within or on the outskirts of cities, the goal is siege: block the entry of food and other necessary supplies into the city and force the hand of the municipal, regional, or even central government. The siege has a particularly storied history in Bolivia, since it was the method of protest used by eighteenth-century Indigenous revolutionaries Túpac Katari and Bartolina Sisas, who organized forty thousand people to lay siege to La Paz for 184 days in 1781. In rural areas, however, the roadblock has tended to focus on highways used to transport economic products from the countryside to trade hubs (Fabricant 2012b).

9. Brent Kaup (2008) unpacks the different ways that resource materiality has affected protests in Bolivia over time, with a focus on the shift from tin (roadblocks) to natural gas (urban shutdowns). Similar research has been conducted around pipelines and conflict in Nigeria (Watts 2008), eastern Europe (Barry 2013), and the Middle East (Mitchell 2011).

10. Colonial histories always shape sociological categories like patronage, making it difficult to identify a garden variety of patronage to which others might be compared. For example, writing about postcolonial Africa, Achille Mbembe (2001, 44) points out that patron-client dynamics have sustained a complex system of revenue transfers from formal circuits to informal ones, from urban to rural, from rich to poor. Even government salaries, Mbembe contends, function more like a gift than ordinary wages for labor since remuneration has nothing to do with actual work.

11. This distinction parallels a more general distinction within the ayllus of Norte Potosí between everyday currency (as controlled by the state) and a more ancient and continuously fertile source of money engendered in the subterranean (Harris 1989).

12. This percentage was listed on receipts from tin commercializers that cooperative miners showed me.

13. Mbembe (2001) describes a scenario in which the relationship between individuals and the state in postcolonial Africa is mediated by extended kinship networks. In his work on the circulation of oil rents in Nigeria, Michael Watts (2001, 2004) similarly documents the mediating scalar effects of kinship groups, regional governments, and even generationally divided organizations.

14. There are nine departments in Bolivia, and in 2016 each of these had only one regional cooperative mining federation except for Potosí, which had three: Central Potosí, Norte Potosí, and Sur Atocha.

15. Similarly, Timothy Mitchell (2011) argues that the physical conditions in which coal miners worked in nineteenth-century Europe created the conditions of possibility for the emergence of democracy, since the occluded passageways and specialized knowledge required to navigate them afforded workers privacy from their bosses—privacy that they used to establish powerful unions.

16. Their names are Freddy Ambrosio, Rubén Arapaya, Severino Ichota, Fermín Mamani, and Pedro Mamani.

17. By 2019 nine miners had been formally convicted of murder and sentenced to six years in prison. Of the convicted, two had been FENCOMIN leaders (*Los Tiempos* 2019).

18. "Actors on all sides of the Bolivian political spectrum use their bodies and charged symbols of Indigeneity, history, and the nation in public performance," writes Nancy Postero (2017, 19). Evo's government has proved particularly adept at mobilizing symbolic power in public spectacles to solidify his own place as rightful leader.

AFTERWORD

1. There is enormous controversy over whether the end of Evo's presidency was caused by electoral fraud or an organized coup. Although I use the word *coup*, I tend to agree with Amy Kennemore and Nancy Postero (2022) and Silvia Rivera Cusicanqui (2020) that the dichotomous framing of this debate—fraud versus coup—obscures many of the complicated dynamics at play and flattens the voices of social movement leaders. Some other analyses of the 2019 crisis from a variety of perspectives can be found in Bjork-James (2021), Farthing and Becker (2021), Gustafson (2020, 247–53), and Shultz (2019).

2. For an exception, see Perreault (2020) for a fabulous analysis of the role of lithium's resource regionalism—and resource politics in general—in shaping the 2019 elections.

3. DS 4721, art. 3.II.i., May 18, 2022, https://www.lexivox.org/norms/BO-DS -N4721.html.

4. I am grateful to Amy Kennemore, who studies Bolivia's plural legal system and who was supporting Tata Félix as he navigated it, for putting us in touch.

5. The problem with the Indigenous Fund, according to activists, was that there was no obvious place to send the money awarded for projects, since most Indigenous communities did not have collective bank accounts. Instead, leaders received the funds directly and were personally tasked with executing projects such as building hospitals and highways—a tall order for anyone who does not have training or contacts in these areas. The projects languished and created a convenient excuse for state persecution.

6. Later in 2013, CONAMAQ leaders were forcibly removed from their offices and replaced with an "official" (state-sponsored) CONAMAQ—dubbed "CONA-MAS" by activists who wanted to make apparent the puppet strings headed to Evo's party, the MAS. The "organic" or original CONAMAQ leadership continued to work out of informal office spaces for a few years before running out of rent money; by the time I spoke with Tata Félix, the organic CONAMAQ was virtually paralyzed by lack of funding.

7. I was not able to record this interview because of prison regulations against technology, so the quotes are approximations derived from furious notebook scribbling.

Abercrombie, Thomas A. 1996. "Q'aqchas and La Plebe in 'Rebellion': Carnival vs. Lent in 18th-Century Potosí." *Journal of Latin American Anthropology* 2 (2): 62–111.

Absi, Pascale. 2005. *Los ministros del diablo: El trabajo y sus representaciones en las minas de Potosí.* La Paz: Institut français d'études andines / Fundación PIEB.

Absi, Pascale. 2006a. "De kajchas a cooperativistas: Historia de una relación con el trabajo, la tierra y el estado." *El Juguete Rabios,* November 11, 2006.

Absi, Pascale. 2006b. "Lifting the Layers of the Mountain's Petticoats: Mining and Gender in Potosi's Pachamama." In *Mining Women: Gender in the Development of a Global Industry, 1670 to 2005,* edited by Jaclyn J. Gier and Laurie Mercier, 58–70. New York: Palgrave Macmillan.

Achtenberg, Emily. 2015. "What's behind the Bolivian Government's Attack on NGOs?" *Rebel Currents* (blog), NACLA, September 3, 2015. https://nacla.org /blog/2015/09/03/what's-behind-bolivian-government's-attack-ngos.

Adas, Michael. (1989) 2015. *Machines as the Measure of Men: Science, Technology, and Ideologies of Western Dominance.* Ithaca, NY: Cornell University Press.

Adey, Peter. 2010. *Aerial Life: Spaces, Mobilities, Affects.* Malden, MA: John Wiley and Sons.

Adler, Antony. 2020. "Deep Horizons: Canada's Underwater Habitat Program and Vertical Dimensions of Marine Sovereignty." *Centaurus* 62 (4): 763–82.

Agard-Jones, Vanessa. 2013. "Bodies in the System." *Small Axe: A Caribbean Journal of Criticism* 17 (3): 182–92.

Agricola, Georgius. (1556) 1950. *De re metallica.* Translated by Herbert Clark Hoover. New York: Dover.

Ahlfeld, Federico. 1931. "The Tin Ores of Uncia-Llallagua, Bolivia." *Economic Geology* 26 (3): 241–57.

Ahlfeld, Federico. 1936. "The Bolivian Tin Belt." *Economic Geology* 31 (1): 48–72.

Ahlfeld, Federico. 1941. "Zoning in the Bolivian Tin Belt." *Economic Geology* 36 (6): 569–88.

Ahlfeld, Federico. (1946) 1972. *Geología de Bolivia.* La Paz: Editorial Los Amigos del Libro.

Akram-Lodhi, A. Haroon, and Cristóbal Kay. 2010a. "Surveying the Agrarian Question (Part 1): Unearthing Foundations, Exploring Diversity." *Journal of Peasant Studies* 37 (1): 177–202.

Akram-Lodhi, A. Haroon, and Cristóbal Kay. 2010b. "Surveying the Agrarian Question (Part 2): Current Debates and Beyond." *Journal of Peasant Studies* 37 (2): 255–84.

Alaimo, Stacy, and Susan Hekman, eds. 2008. *Material Feminisms*. Bloomington: Indiana University Press.

Albarracín Millán, Juan. 2002. *Una visión esplendorosa de Bolivia: Las exploraciones de Alcides d'Orbigny en Bolivia*. La Paz: Plural Editores.

Albro, Robert. 2006. "The Culture of Democracy and Bolivia's Indigenous Movements." *Critique of Anthropology* 26 (4): 387–410.

Alves, Andrés A., and José M. Moreira. 2013. *The Salamanca School*. London: Bloomsbury.

Anderson, Kay. 2007. *Race and the Crisis of Humanism*. New York: Routledge.

Andolina, Robert, Sarah Radcliffe, and Nina Laurie. 2005. "Development and Culture: Transnational Identity Making in Bolivia." *Political Geography* 24 (6): 678–702.

Andrew, Edward. 1983. "Class in Itself and Class against Capital: Karl Marx and his Classifiers." *Canadian Journal of Political Science / Revue canadienne de science politique* 16 (3): 577–84.

Anthias, Floya, and Nina Yuval-Davis. 2005. *Racialized Boundaries: Race, Nation, Gender, Colour and Class and the Anti-racist Struggle*. London: Routledge.

Anthias, Penelope. 2018. *Limits to Decolonization: Indigeneity, Territory, and Hydrocarbon Politics in the Bolivian Chaco*. Ithaca, NY: Cornell University Press.

Anthony, Patrick. 2018. "Mining as the Working World of Alexander von Humboldt's Plant Geography and Vertical Cartography." *Isis* 109 (1): 28–55.

Arbona, Juan Manuel. 2007. "Neo-liberal Ruptures: Local Political Entities and Neighbourhood Networks in El Alto, Bolivia." *Geoforum* 38 (1): 127–37.

Arce, Roberto. 1965. *Recomendaciones para la rehabilitación de la industria minera*. La Paz: Don Bosco.

Aresti, María Lasa. 2016. *Oil and Gas Revenue Sharing in Bolivia*. New York: Natural Resource Governance Institute.

Arguedas, Alcides. (1909) 2008. *Pueblo enfermo*. La Paz: Libreria Editorial "G.U.M."

Arze Vargas, Carlos. 2021. "El jukeo y la política minera masista." CEDLA: *Reporte Annual de Industrias Extractivas* 6:148–82.

Auyero, Javier. 2000. "The Logic of Clientelism in Argentina: An Ethnographic Account." *Latin American Research Review* 35 (3): 55–81.

Ford, Bacon, and Davis. 1956. *Mining Industry of Bolivia: A Report to the Bolivian Ministry of Mines and Petroleum*. New York: Ford, Bacon, and Davis.

Baker, Ian. 2018. "Tin." In *Fifty Materials That Make the World*, 235–39. Cham, Switzerland: Springer.

Bakewell, Peter. 1984. *Miners of the Red Mountain: Indian Labor in Potosí, 1545–1650*. Albuquerque: University of New Mexico Press.

Bakker, Karen J. 2003. "A Political Ecology of Water Privatization." *Studies in Political Economy* 70 (1): 35–58.

Bakker, Karen, and Gavin Bridge. 2006. "Material World? Resource Geographies and the 'Matter of Nature.'" *Progress in Human Geography* 30 (1): 5–27.

Bakker, Karen, and Gavin Bridge. 2021. "Material Worlds Redux: Mobilizing Materiality within Critical Resource Geography." In *The Routledge Handbook of Critical Resource Geography*, edited by Matthew Himley, Elizabeth Havice, and Gabriela Valdivia, 43–56. London: Routledge.

Ballestero, Andrea. 2019. *A Future History of Water*. Durham, NC: Duke University Press.

Barad, Karen. 2007. *Meeting the Universe Halfway: Quantum Physics and the Entanglement of Matter and Meaning*. Durham, NC: Duke University Press.

Barragán Romano, Rossana. 2015. "¿Ladrones, pequeños empresarios o trabajadores independientes? K'ajchas, trapiches y plata en el cerro de Potosí en el siglo XVIII." *Nuevo Mundo, Mundos Nuevos*, March 10, 2015. https://doi.org/10.4000/nuevomundo.67938.

Barragán Romano, Rossana. 2017. "Los k'ajchas* y los proyectos de industria y nación en Bolivia (1935–1940)." *Revista Mundos do Trabalho* 9 (18): 25–48.

Barry, Andrew. 2013. *Material Politics: Disputes along the Pipeline*. Oxford: Wiley-Blackwell.

Bebbington, Anthony, and Jeffrey Bury, eds. 2013. *Subterranean Struggles: New Dynamics of Mining, Oil, and Gas in Latin America*. Austin: University of Texas Press.

Benjamin, Walter. (1968) 2020. "Theses on the Philosophy of History." In *Critical Theory and Society: A Reader*, edited by Stephen Eric Bronner and Douglas MacKay Kellner, 255–63. New York: Routledge.

Bennett, Jane. 2010. *Vibrant Matter: A Political Ecology of Things*. Durham, NC: Duke University Press.

Béraud, Gilles. 2000. "Le Départ d'Alcide d'Orbigny pour l'Amérique Méridionale et sa Préparation." *Annales de la Société des Sciences Naturelles de la Charente-Maritime* 8 (9): 1118–28.

Berlant, Lauren. 2011. *Cruel Optimism*. Durham, NC: Duke University Press.

Bernstein, Henry. 1996. "Agrarian Questions Then and Now." *Journal of Peasant Studies* 24 (1–2): 22–59.

Billé, Franck, ed. 2019. "Volumetric Sovereignty: A Forum." *Environment and Planning D: Society and Space*, April 10, 2019. http://societyandspace.org/2019/04/10/volumetricsovereigntyforum/.

Billé, Franck, ed. 2020. *Voluminous States: Sovereignty, Materiality, and the Territorial Imagination*. Durham, NC: Duke University Press.

Bjork-James, Carwill. 2020. *The Sovereign Street: Making Revolution in Urban Bolivia*. Tucson: University of Arizona Press.

Bjork-James, Carwill. 2021. "'We Are MAS 2.0': Returning to Power a Year after the Coup against Evo Morales, Bolivia's MAS Party Must Tackle Steep Challenges While Rebuilding Relationships with Its Grassroots Base." *NACLA Report on the Americas* 53 (1): 7–11.

Blanco, Hugo. 1972. *Land or Death: The Peasant Struggle in Peru*. New York: Pathfinder.

Bobbette, Adam, and Amy Donovan. 2019. *Political Geology: Active Stratigraphies and the Making of Life*. Cham, Switzerland: Palgrave Macmillan.

Bosworth, Kai. 2017. "Thinking Permeable Matter through Feminist Geophilosophy: Environmental Knowledge Controversy and the Materiality of Hydrogeologic Processes." *Environment and Planning D: Society and Space* 35 (1): 21–37.

Braun, Bruce. 2000. "Producing Vertical Territory: Geology and Governmentality in Late Victorian Canada." *Cultural Geographies* 7 (1): 7–46.

Braun, Bruce. 2002. *The Intemperate Rainforest: Nature, Culture, and Power on Canada's West Coast*. Minneapolis: University of Minnesota Press.

Braun, Bruce, and Sarah J. Whatmore, eds. 2010. *Political Matter: Technoscience, Democracy, and Public Life*. Minneapolis: University of Minnesota Press.

Brenner, Neil, and Stuart Elden. 2009. "Henri Lefebvre on State, Space, Territory." *International Political Sociology* 3 (4): 353–77.

Bridge, Gavin. 2013. "Territory, Now in 3D!" *Political Geography* 34 (C): 55–57.

Brown, Jacqueline Nassy. 2005. *Dropping Anchor, Setting Sail: Geographies of Race in Black Liverpool*. Princeton, NJ: Princeton University Press.

Budds, Jessica, and Leonith Hinojosa. 2012. "Restructuring and Rescaling Water Governance in Mining Contexts: The Co-production of Waterscapes in Peru." *Water Alternatives* 5 (1): 119–37.

Burke, Melvin. 1987. "The Corporación Minera de Bolivia (COMIBOL) and the Triangular Plan: A Case Study in Dependency." *Latin American Issues* 4. http://sites.allegheny.edu/latinamericanstudies/latin-american-issues/volume-4/.

Burman, Anders. 2014. "'Now We Are Indígenas': Hegemony and Indigeneity in the Bolivian Andes." *Latin American and Caribbean Ethnic Studies* 9 (3): 247–71.

Burman, Anders. 2015. "El ayllu y el indianismo: Autenticidad, representatividad y territorio en el quehacer político del CONAMAQ, Bolivia." In *Los nuevos caminos de los movimientos sociales en Latinoamérica*, edited by Anne Marie Ejdesgaard Jeppsen, Helene Balslev Clausen, and Mario Alberto Velázquez García, 100–122. Monterrey: Tilde Editores.

Buss, Doris, Blair Rutherford, Jennifer Stewart, Gisèle Eva Côté, Abby Sebina-Zziwa, Richard Kibombo, Jennifer Hinton, and Joanne Lebert. 2019. "Gender and Artisanal and Small-Scale Mining: Implications for Formalization." *Extractive Industries and Society* 6 (4): 1101–12.

Butler, Judith. 1993. *Bodies That Matter: On the Discursive Limits of "Sex."* New York: Routledge.

Callon, Michel. 1984. "Some Elements of a Sociology of Translation: Domestication of the Scallops and the Fishermen of St. Brieuc Bay." *Sociological Review* 32 (1): 196–233.

Campbell, Elaine. 2019. "Three-Dimensional Security: Layers, Spheres, Volumes, Milieus." *Political Geography* 69:10–21.

Canaviri Sirpa, Teresa, and Roxana Selum Yaveta. 2005. *Warmi mineral y copajira: Memoria de mujeres en diez cooperativas*. La Paz: Viceministro de la Mujer.

Canelas Orellana, Amada. 1981. *¿Quiebra de la minería estatal boliviana?* La Paz: Los Amigos del Libro.

Canessa, Andrew. 2007. "Who Is Indigenous? Self-Identification, Indigeneity, and Claims to Justice in Contemporary Bolivia." *Urban Anthropology and Studies of Cultural Systems and World Economic Development* 36 (3): 195–237.

Canessa, Andrew. 2012. *Intimate Indigeneities: Race, Sex, and History in the Small Spaces of Andean Life.* Durham, NC: Duke University Press.

Canessa, Andrew. 2014. "Conflict, Claim and Contradiction in the New 'Indigenous' State of Bolivia." *Critique of Anthropology* 34 (2): 153–73.

Cannon, Susan Faye. 1978. *Science in Culture: The Early Victorian Period.* New York: Science History Publications.

Carrasco, Gloria. 2017. "Condenan a 10 años de cárcel a Gabriela Zapata, expareja de Evo Morales." CNN Latinoamérica, May 23, 2017.

Castree, Noel. 2000. "Marxism and the Production of Nature." *Capital and Class* 24 (3): 5–36.

Céspedes, Augusto. (1946) 1986. *Metal del diablo.* La Paz: Libreria Editorial Juventud.

Chayanov, Alexander. (1927) 1991. *The Theory of Peasant Co-operatives.* Translated by David Wedgwood Benn. Columbus: Ohio State University Press.

Chen, Mel Y. 2012. *Animacies: Biopolitics, Racial Mattering, and Queer Affect.* Durham, NC: Duke University Press.

Choque, María Eugenia, and Carlos Mamani. 2001. "Reconstitución del ayllu y derechos de los pueblos indígenas: El movimiento indio en los Andes de Bolivia." *Journal of Latin American Anthropology* 6 (1): 202–24.

Clark, Nigel. 2011. *Inhuman Nature: Sociable Life on a Dynamic Planet.* London: Sage.

Clark, Nigel, and Kathryn Yusoff. 2017. "Geosocial Formations and the Anthropocene." *Theory, Culture and Society* 34 (2–3): 3–23.

Cleary, David. 1990. *Anatomy of the Amazon Gold Rush.* London: Palgrave Macmillan.

Colloredo-Mansfeld, Rudi. 1998. "'Dirty Indians,' Radical Indígenas, and the Political Economy of Social Difference in Modern Ecuador. *Bulletin of Latin American Research* 17 (2): 185-205.

Coronil, Fernando. 1997. *The Magical State: Nature, Money, and Modernity in Venezuela.* Chicago: University of Chicago Press.

Coulthard, Glen Sean. 2014. *Red Skin, White Masks: Rejecting the Colonial Politics of Recognition.* Minneapolis: University of Minnesota Press.

Coyle Rosen, Lauren. 2020. *Fires of Gold: Law, Spirit, and Sacrificial Labor in Ghana.* Berkeley: University of California Press.

CPE (Constitución Política del Estado Plurinacional de Bolivia). February 7, 2009. https://www.oas.org/dil/esp/Constitucion_Bolivia.pdf.

Crenshaw, Kimberlé Williams. 1991. "Mapping the Margins: Intersectionality, Identity Politics, and Violence against Women of Color." *Stanford Law Review* 43 (6): 1241–99.

Crowfoot, Arthur. 1914. "Development of the Round Table at Great Falls." *Bulletin of the American Institute of Mining Engineers* no. 92 (August 1914): 1931–83.

D'Angelo, Lorenzo. 2014. "Who Owns the Diamonds? The Occult Economy of Diamond Mining in Sierra Leone." *Africa* 84 (2): 269–93.

d'Avignon, Robyn. 2018. "Primitive Techniques: From 'Customary' to 'Artisanal' Mining in French West Africa." *Journal of African History* 59 (2): 179–97.

Deister Concentrator Company. 1906. *The Deister Concentrator: Especially Adapted for the Treatment of Fines and Slimes.* Fort Wayne, IN: Deister Concentrator Company. Accessed January 14, 2019. https://web.archive.org/web/20190115233025/https://www.deisterconcentrator.com/pdf/deister_book.pdf.

de la Cadena, Marisol. 2000. *Indigenous Mestizos: The Politics of Race and Culture in Cuzco, Peru, 1919–1991.* Durham, NC: Duke University Press.

de la Cadena, Marisol. 2005. "Are 'Mestizos' Hybrids? The Conceptual Politics of Andean Identities." *Journal of Latin American Studies* 37 (2): 259–84.

de la Cadena, Marisol. 2010. "Indigenous Cosmopolitics in the Andes: Conceptual Reflections beyond 'Politics.'" *Cultural Anthropology* 25 (2): 334–70.

Demeritt, David. 2002. "What Is the 'Social Construction of Nature'? A Typology and Sympathetic Critique." *Progress in Human Geography* 26 (6): 767–90.

de Theije, Marjo. 2020. "Brazil: Forever Informal." In *Global Gold Production Touching Ground: Expansion, Informalization, and Technological Innovation*, edited by Boris Verbrugge and Sara Geenen, 117–34. Cham, Switzerland: Palgrave Macmillan.

Díaz Romero, Belisario. 1907. Introduction to *Estudios sobre la geología de Bolivia*, by Alcides Dessalines d'Orbigny and Victor E. Marchant Y., i–xix. La Paz: Ministerio de Colonización y Agricultura.

d'Orbigny, Alcide Dessalines. 1835–47. *Voyage dans l'Amérique méridionale.* Paris: Société Géologique de France.

d'Orbigny, Alcide Dessalines. 1839. *L'homme américain (de l'Amérique méridionale) consideré sous ses rapports physiologiques et moraux.* Paris: Pitois-Levrault et C., Libraires-Éditeurs.

d'Orbigny, Alcide Dessalines. 1845. *Descripción geográfica, histórica y estadística de Bolivia.* Paris: Librería de los Señores Gide y Compañía.

d'Orbigny, Alcides Dessalines. 1944. *El hombre americano considerado en sus aspectos fisiológicos y morales.* Buenos Aires: Editorial Futuro.

d'Orbigny, Alcides Dessalines. 1946. *Descripción geográfica, histórica y estadística de Bolivia.* La Paz: Instituto Cultural Anglo-Boliviano.

d'Orbigny, Alcides Dessalines, and Victor E. Marchant Y. 1907. *Estudios sobre la geología de Bolivia.* La Paz: Ministerio de Colonización y Agricultura.

d'Orbigny, Alcide, and Andrés de Santa Cruz. (1830) 2017. "Correspondance entre Alcide d'Orbigny et le Maréchal Andrés de Santa Cruz." Edited by Pierre-Olivier Combelles. *Rouge et Blanc, ou le Fil d'Ariane d'un voyageur naturaliste* (blog), December 11, 2017. http://pocombelles.over-blog.com/article-14546919.html.

Douglas, Mary. (1966) 2003. *Purity and Danger: An Analysis of Concepts of Pollution and Taboo.* New York: Routledge.

Duffield, J. R., C. R. Morris, D. M. Morriah, J. A. Vesey, and D. R. Williams. 1989. "The Speciation and Bioavailability of Tin in Biofuels." In *Tin-Based Antitumour Drugs*, edited by Marcel Gielen, 147–69. New York: NATO Advanced Science Affairs Division.

Dunkerley, James. 1984. *Rebellion in the Veins: Political Struggle in Bolivia, 1952–1982*. London: Verso.

Elden, Stuart. 2013. "Secure the Volume: Vertical Geopolitics and the Depth of Power." *Political Geography* 34 (C): 35–51.

Ellison, Susan Helen. 2018. *Domesticating Democracy: The Politics of Conflict Resolution in Bolivia*. Durham, NC: Duke University Press.

Emel, Jody, Matthew T. Huber, and Madoshi H. Makene. 2011. "Extracting Sovereignty: Capital, Territory, and Gold Mining in Tanzania." *Political Geography* 30 (2): 70–79.

Escobar, Filemón. 1986. *La mina vista desde el guardatojo*. La Paz: CIPCA.

Escóbar de Pabón, Silvia, Bruno Rojas Callejas, and Giovanna Hurtado Aponte. 2016. *Jóvenes asalariados y precariedad laboral: Situación de los derechos laborales en Bolivia 2012-2015*. La Paz: CEDLA.

Espinoza Morales, Jorge. 2010. *Minería boliviana: Su realidad*. La Paz: Plural Editores.

Espinoza Morales, Jorge. 2013. "Minería estatal: Una historia de fracasos?" In *De vuelta al estado minero?*, edited by Henry Oporto, Jorge Espinoza, Rubén Ferrufino, Dionisio Garzón, and Héctor E. Córdova, 49–126. La Paz: Fundación Vicente Pazos Kanki.

Fabricant, Nicole. 2012a. *Mobilizing Bolivia's Displaced: Indigenous Politics and the Struggle over Land*. Chapel Hill: University of North Carolina Press.

Fabricant, Nicole. 2012b. "Symbols in Motion: Katari as Traveling Image in Landless Movement Politics in Bolivia." *Latin American and Caribbean Ethnic Studies* 7 (1): 1–29.

Fabricant, Nicole, and Nancy Postero. 2015. "Sacrificing Indigenous Bodies and Lands: The Political-Economic History of Lowland Bolivia in Light of the Recent TIPNIS Debate." *Journal of Latin American and Caribbean Anthropology* 20 (3): 452–74.

Farthing, Linda, and Thomas Becker. 2021. *Coup: A Story of Violence and Resistance in Bolivia*. Chicago: Haymarket.

FENCOMIN. 2008. *Recopilación histórica de congresos*. La Paz: Federación Nacional de Cooperativas Mineras de Bolivia.

Ferreira, Javo. 2010. *Comunidad, indigenismo y marxismo: Un debate sobre la cuestión agraria y nacional-indígena en los Andes*. La Paz: Ediciones Palabra Obrera.

Ferry, Elizabeth Emma. 2002. "Inalienable Commodities: The Production and Circulation of Silver and Patrimony in a Mexican Mining Cooperative." *Cultural Anthropology* 17 (3): 331–58.

Ferry, Elizabeth Emma. 2005. *Not Ours Alone: Patrimony, Value, and Collectivity in Contemporary Mexico*. New York: Columbia University Press.

Field, Thomas C., Jr. 2014. *From Development to Dictatorship: Bolivia and the Alliance for Progress in the Kennedy Era*. Ithaca, NY: Cornell University Press.

Foucault, Michel. (1976) 2003. *"Society Must Be Defended": Lectures at the Collège de France, 1975–1976.* New York: Macmillan.

Francescone, Kirsten. 2015. "Cooperative Miners and the Politics of Abandonment in Bolivia." *Extractive Industries and Society* 2 (4): 746–55.

Francescone, Kirsten, and Vladimir Díaz. 2013. "Cooperativas mineras: Entre socios, patrones y peones." *Petropress* 30 (1): 32–41.

Frank, Andre Gunder. 1972. *Lumpenbourgeoisie: Lumpendevelopment. Dependence, Class, and Politics in Latin America.* New York: Monthly Review.

Freitas, Carlos Machado, and Mariano Andrade da Silva. 2019. "Work Accidents Which Become Disasters: Mine Tailing Dam Failures in Brazil." *Revista Brasileira de Medicina do Trabalho* 17 (1): 21–29.

French, Jan Hoffman. 2009. *Legalizing Identities: Becoming Black or Indian in Brazil's Northeast.* Chapel Hill: University of North Carolina Press.

Galeano, Eduardo. (1973) 1997. *Open Veins of Latin America.* New York: Monthly Review.

Galindo, María. 2017. *No hay libertad política si no hay libertad sexual.* La Paz: Mujeres Creando.

García Linera, Álvaro. 2014. *Identidad boliviana: Nación, mestizaje, y plurinacionalidad.* La Paz: Vicepresidencia del Estado Plurinacional.

Geddes, Charles F. 1972. *Patiño: El rey del estaño.* La Paz: Editor El Amigo.

Gerlofs, Ben A. 2021. "Seismic Shifts: Recentering Geology and Politics in the Anthropocene." *Annals of the American Association of Geographers* 111 (3): 828–36.

Gifford, Terry. 1996. "The Social Construction of Nature." *Interdisciplinary Studies in Literature and Environment* 3 (2): 27–35.

Gill, Lesley. 1997. "Relocating Class: Ex-miners and Neoliberalism in Bolivia." *Critique of Anthropology* 17 (3): 293–312.

Gill, Lesley. 2000. *Teetering on the Rim: Global Restructuring, Daily Life, and the Armed Retreat of the Bolivian State.* New York: Columbia University Press.

Godoy, Ricardo A. 1985a. "State, Ayllu, and Ethnicity in Northern Potosí, Bolivia." *Anthropos* 80 (1/3): 53–65.

Godoy, Ricardo A. 1985b. "Technical and Economic Efficiency of Peasant Miners in Bolivia." *Economic Development and Cultural Change* 34 (1): 103–20.

Goldberg, David Theo. 2002. *The Racial State.* Malden, MA: Blackwell.

Goldstein, Daniel M. 2016. *Owners of the Sidewalk: Security and Survival in the Informal City.* Durham, NC: Duke University Press.

Gordillo, Gaston R. 2014. *Rubble: The Afterlife of Destruction.* Durham, NC: Duke University Press.

Gotkowitz, Laura. 2007. *A Revolution for Our Rights: Indigenous Struggles for Land and Justice in Bolivia, 1880–1952.* Durham, NC: Duke University Press.

Gowan, Peter. 1999. *The Global Gamble: Washington's Faustian Bid for World Dominance.* New York: Verso.

Graham, Richard, ed. 1990. *The Idea of Race in Latin America, 1870–1940.* Austin: University of Texas Press.

Graham, Stephen. 2016. *Vertical: The City from Satellites to Bunkers*. New York: Verso.

Graulau, Jeannette. 2011. "Ownership of Mines and Taxation in Castilian Laws, from the Middle Ages to the Early Modern Period: The Decisive Influence of the Sovereign in the History of Mining." *Continuity and Change* 26 (1): 13–44.

Grisaffi, Thomas. 2018. *Coca Yes, Cocaine No: How Bolivia's Coca Growers Reshaped Democracy*. Durham, NC: Duke University Press.

Grosz, Elizabeth A. 2008. *Chaos, Territory, Art: Deleuze and the Framing of the Earth*. New York: Columbia University Press.

Gudynas, Eduardo. 2009. "Diez tesis urgentes sobre el nuevo extractivismo: Contextos y demandas bajo el progresismo sudamericano actual." In *Extractivismo, política y sociedad*, edited by Francisco Rhon Dávila, 187–225. Quito: CAAP/CLAES.

Guldi, Jo. 2022. *The Long Land War: The Global Struggle for Occupancy Rights*. New Haven, CT: Yale University Press.

Gustafson, Bret. 2009a. "Manipulating Cartographies: Plurinationalism, Autonomy, and Indigenous Resurgence in Bolivia." *Anthropological Quarterly* 82 (4): 985–1016.

Gustafson, Bret. 2009b. *New Languages of the State: Indigenous Resurgence and the Politics of Knowledge in Bolivia*. Durham, NC: Duke University Press.

Gustafson, Bret. 2020. *Bolivia in the Age of Gas*. Durham, NC: Duke University Press.

Guthman, Julie. 2011. *Weighing In: Obesity, Food Justice, and the Limits of Capitalism*. Berkeley: University of California Press.

Haggard, Henry Rider. 1885. *King Solomon's Mines*. London: Dent.

Hale, Charles R. 2002. "Does Multiculturalism Menace? Governance, Cultural Rights and the Politics of Identity in Guatemala." *Journal of Latin American Studies* 34 (3): 485–524.

Hale, Charles R. 2004. "Rethinking Indigenous Politics in the Era of the 'Indio Permitido.'" *NACLA Report on the Americas* 38 (2): 16–21.

Hall, Stuart. 2003. "Marx's Notes on Method: A 'Reading' of the '1857 Introduction.'" *Cultural Studies* 17 (2): 113–49.

Hanke, Lewis. 1974. *All Mankind Is One: A Study of the Disputation between Bartolomé de Las Casas and Juan Ginés de Sepúlveda on the Religious and Intellectual Capacity of the American Indians*. Dekalb: Northern Illinois University Press.

Haraway, Donna. 1988. "Situated Knowledges: The Science Question in Feminism and the Privilege of Partial Perspective." *Feminist Studies* 14 (3): 575–99.

Harding, Sandra. 1992. "After the Neutrality Ideal: Science, Politics, and 'Strong Objectivity.'" *Social Research* 59 (3): 567–87.

Harris, Olivia. 1989. "The Earth and the State: The Sources and Meanings of Money in Northern Potosi, Bolivia." In *Money and the Morality of Exchange*, edited by Jonathan Parry and Maurice Bloch, 232–68. Cambridge: Cambridge University Press.

Harris, Olivia. 1995. "Ethnic Identities and Market Relations: Indian and Mestizos in the Andes." In *Ethnicity, Markets, and Migration in the Andes: At the Crossroads of Anthropology and History*, edited by Olivia Harris and Brooke Larson, 351–90. Durham, NC: Duke University Press.

Harris, Olivia, and Xavier Albó. 1976. *Monteras y guardatojos: Campesinos y mineros en el norte de Potosí*. La Paz: CIPCA.

Hart, Gillian. 2018. "Relational Comparison Revisited: Marxist Postcolonial Geographies in Practice." *Progress in Human Geography* 42 (3): 371–94.

Hartman, Saidiya. 2019. *Wayward Lives, Beautiful Experiments: Intimate Histories of Riotous Black Girls, Troublesome Women, and Queer Radicals*. New York: W. W. Norton.

Hartsock, Nancy, and Neil Smith. 1979. "On Althusser's Misreading of Marx's 1857 'Introduction.'" *Science and Society* 43 (4): 486–89.

Harvey, David. 1982. *The Limits to Capital*. New York: Verso.

Harvey, David. 2001. "Globalization and the 'Spatial Fix.'" *Geographische Revue* 3 (2): 23–30.

Hawkins, Harriet. 2020. "'A Volcanic Incident': Toward a Geopolitical Aesthetics of the Subterranean." *Geopolitics* 25 (1): 214–39.

Hegel, Georg Wilhelm Friedrich. (1822–28) 1997. "Lectures on the Philosophy of World History." In *Race and the Enlightenment: A Reader*, compiled and edited by Emmanuel Chukwudi Eze, 109–49. Cambridge, MA: Blackwell.

Heilbroner, Robert L. 1967. "Do Machines Make History?" *Technology and Culture* 8 (3): 335–45.

Heynen, Nik, James McCarthy, Scott Prudham, and Paul Robbins, eds. 2007. *Neoliberal Environments: False Promises and Unnatural Consequences*. New York: Routledge.

High, Mette M. 2017. *Fear and Fortune: Spirit Worlds and Emerging Economies in the Mongolian Gold Rush*. Ithaca, NY: Cornell University Press.

Hilson, Gavin. 2016. "Farming, Small-Scale Mining and Rural Livelihoods in Sub-Saharan Africa: A Critical Overview." *Extractive Industries and Society* 3 (2): 547–63.

Himley, Matthew. 2021. "The Future Lies Beneath: Mineral Science, Resource-Making, and the (De)Differentiation of the Peruvian Underground." *Political Geography* 87: 102373.

Hindery, Derrick. 2013. *From Enron to Evo: Pipeline Politics, Global Environmentalism, and Indigenous Rights in Bolivia*. Tucson: University of Arizona Press.

Hobsbawm, E. J. 1973. "Peasants and Politics." *Journal of Peasant Studies* 1 (1): 3–22.

Hooker, Juliet. 2005. "'Beloved Enemies': Race and Official Mestizo Nationalism in Nicaragua." *Latin American Research Review* 40 (3): 14–39.

Howard, April, and Benjamin Dangl. 2006. "Tin War in Bolivia: Conflict between Miners Leaves 17 Dead." *Upside Down World*, October 10, 2006. https://upsidedownworld.org/archives/bolivia/tin-war-in-bolivia-conflict-between-miners-leaves-17-dead/.

Humboldt, Alexander von, and Aimé Bonpland. 1807. *Essai sur la géographie des plantes, accompagné d'un tableau physique des regions équinoxiales*. Paris: F. Schoell.

Hummel, Calla, Felicia Marie Knaul, Michael Touchton, V. Ximena Velasco Guachalla, Jami Nelson-Nuñez, and Carew Boulding. 2021. "Poverty, Precarious Work, and the COVID-19 Pandemic: Lessons from Bolivia." *Lancet Global Health* 9 (5): e579–81.

Humphreys Bebbington, Denise, and Anthony J. Bebbington. 2010. "Extraction, Territory, and Inequalities: Gas in the Bolivian Chaco." *Canadian Journal of Development Studies / Revue canadienne d'études du développement* 30 (1–2): 259–80.

Ingulstad, Mats, Andrew Perchard, and Espen Storli, eds. 2015. *Tin and Global Capitalism: A History of the Devil's Metal, 1850–2000*. New York: Routledge.

Intergovernmental Forum on Mining, Minerals and Sustainable Development. 2017. *Global Trends in Artisanal and Small-Scale Mining (ASM): A Review of Key Numbers and Issues*. Winnipeg: International Institute for Sustainable Development.

International Monetary Fund. 2014. *Staff Report for the 2013 Article IV Consultation: IMF Country Report No. 14/36*. Washington, DC: International Monetary Fund.

International Tin Association. 2020. Artisanal and Small-Scale Mining Policy. February 2020. https://www.internationaltin.org/artisanal-small-scale-mining/.

Jackson, John P., and Nadine M. Weidman. 2004. *Race, Racism, and Science: Social Impact and Interaction*. Santa Barbara, CA: ABC-CLIO.

Jackson, Zakiyyah Iman. 2020. *Becoming Human: Matter and Meaning in an Antiblack World*. New York: New York University Press.

Jaramillo, Pablo. 2020. "Mining Leftovers: Making Futures on the Margins of Capitalism." *Cultural Anthropology* 35 (1): 48–73.

Jasanoff, Sheila. 2015. "Future Imperfect: Science, Technology, and the Imaginations of Modernity." In *Dreamscapes of Modernity: Sociotechnical Imaginaries and the Fabrication of Power*, edited by Sheila Jasanoff and Sang-Hyun Kim, 1–33. Chicago: University of Chicago Press.

John, S. Sándor. 2009. *Bolivia's Radical Tradition: Permanent Revolution in the Andes*. Tucson: University of Arizona Press.

Johnson, D. Barrie, and Kevin B. Hallberg. 2005. "Acid Mine Drainage Remediation Options: A Review." *Science of the Total Environment* 338 (1–2): 3–14.

Jue, Melody. 2020. *Wild Blue Media: Thinking through Seawater*. Durham, NC: Duke University Press.

Kaup, Brent Z. 2008. "Negotiating through Nature: The Resistant Materiality and Materiality of Resistance in Bolivia's Natural Gas Sector." *Geoforum* 39 (5): 1734–42.

Kaup, Brent Z. 2010. "A Neoliberal Nationalization? The Constraints on Natural-Gas-Led Development in Bolivia." *Latin American Perspectives* 37 (3): 123–38.

Kemp, Deanna, Carol J. Bond, Daniel M. Franks, and Claire Cote. 2010. "Mining, Water and Human Rights: Making the Connection." *Journal of Cleaner Production* 18 (15): 1553–62.

Kennemore, Amy. 2020. "The Search for Indigenous Justice in Plurinational Bolivia: Contested Sovereignties, Entanglement, and the Politics of Harm." PhD diss., University of California, San Diego.

Kennemore, Amy, and Nancy Postero. 2022. "Making Sense of the 2019 Electoral Crisis in Bolivia: Lessons from Indigenous Social Movements." *Foro Internacional* 62 (4): 877–99.

Kiilsgaard, Thor H., Harry A. Tourtelot, Norman J. Page, and S. J. Kropschot. 1992. "Background Notes on U.S. Geological Survey-Bolivia Cooperative Activities." In *Geology and Mineral Resources of the Altiplano and Cordillera Occidental, Bolivia*, 1–3. Denver: US Geological Survey.

Kinchy, Abby J., Roopali Phadke, and Jessica M. Smith. 2018. "Engaging the Underground: An STS Field in Formation." *Engaging Science, Technology, and Society* 4:22–42.

Klauser, Francisco. 2021. "Police Drones and the Air: Toward a Volumetric Geopolitics of Security." *Swiss Political Science Review* 27 (1): 158–69.

Klein, Herbert S. 2003. *A Concise History of Bolivia*. Cambridge: Cambridge University Press.

Klein, Naomi. 2007. *The Shock Doctrine: The Rise of Disaster Capitalism*. New York: Allen Lane.

Klinger, Julie Michelle. 2018. *Rare Earth Frontiers: From Terrestrial Subsoils to Lunar Landscapes*. Ithaca, NY: Cornell University Press.

Klinke, Ian. 2021. "On the History of a Subterranean Geopolitics." *Geoforum* 127:356–63.

Kohl, Benjamin H., and Linda C. Farthing. 2006. *Impasse in Bolivia: Neoliberal Hegemony and Popular Resistance*. London: Zed.

Kohl, Benjamin H., and Linda C. Farthing. 2012. "Material Constraints to Popular Imaginaries: The Extractive Economy and Resource Nationalism in Bolivia." *Political Geography* 31 (4): 225–35.

Kohl, Benjamin H., Linda C. Farthing, and Félix Muruchi. 2011. *From the Mines to the Streets: A Bolivian Activist's Life*. Austin: University of Texas Press.

Lahiri-Dutt, Kuntala. 2018. *Between the Plough and the Pick: Informal, Artisanal and Small-Scale Mining in the Contemporary World*. Acton: Australian National University Press.

Laing, Anna F. 2015. "Resource Sovereignties in Bolivia: Re-conceptualising the Relationship between Indigenous Identities and the Environment during the TIPNIS Conflict." *Bulletin of Latin American Research* 34 (2): 149–66.

La Patria. 2010. "Concluye el proyecto APEMIN II pero se crearán dos sustitutos." *La Patria*, June 19. https://impresa.lapatria.bo/noticia/31705/concluye-el-proyecto-apemin-ii-pero-se-crearan-dos-sustitutos.

Larson, Brooke. 2004. *Trials of Nation Making: Liberalism, Race, and Ethnicity in the Andes, 1810–1910*. Cambridge: Cambridge University Press.

Laserna, Roberto. (2005) 2011. *La trampa del rentismo . . . y cómo salir de ella*. La Paz: Fundación Milenio.

Lave, Rebecca, Matthew W. Wilson, Elizabeth S. Barron, Christine Biermann, Mark A. Carey, Chris S. Duvall, Leigh Johnson, et al. 2014. "Intervention: Critical Physical Geography." *Canadian Geographer / Le Géographe canadien* 58 (1): 1–10.

Lazar, Sian. 2008. *El Alto, Rebel City: Self and Citizenship in Andean Bolivia*. Durham, NC: Duke University Press.

Le Billon, Philippe. 2001. "The Political Ecology of War: Natural Resources and Armed Conflicts." *Political Geography* 20 (5): 561–84.

Lefebvre, Henri. (1974) 1991. *The Production of Space*. Translated by Donald Nicholson-Smith. Oxford: Blackwell.

Le Gouill, Claude. 2014. "La otra cara del katarismo: La experiencia katarista de los ayllus del Norte Potosí." *T'inkazos* 35 (17): 95–113.

Lehm A., Zulema, and Silvia Rivera Cusicanqui. 2005. "A Brief History of Anarchism in Bolivia." *Perspectives on Anarchist History* 9 (1): 24–31.

Libassi, Matthew. 2020. "Mining Heterogeneity: Diverse Labor Arrangements in an Indonesian Informal Gold Economy." *Extractive Industries and Society* 7 (3): 1036–45.

Libassi, Matthew. 2022. "Contested Subterranean Territory: Gold Mining and Competing Claims to Indonesia's Underground." *Political Geography* 98: 102675.

López Vigil, José Ignacio. 1985. *Una mina de coraje: Radios mineras de Bolivia*. Quito: ALER-PÍO XII.

Lora, Guillermo. (1946) 2011. *Tesis de Pulacayo*. Marxists Internet Archive, February 2011. https://www.marxists.org/espanol/lora/1946/nov08.htm.

Lora, Guillermo. 1980. *Historia del movimiento obrero boliviano 1933–1952*. La Paz: Editorial Los Amigos del Libro.

Los Tiempos. 2019. "Caso Illanes: Condenan a 6 años de prisión a expresidente de FENCOMIN y a 4 mineros." *Los Tiempos*, October 18, 2019.

Lottermoser, Bernd G. 2010. *Mine Wastes: Characterization, Treatment, and Environmental Impacts*. New York: Springer.

Luning, Sabine, and Robert J. Pijpers. 2017. "Governing Access to Gold in Ghana: In-Depth Geopolitics on Mining Concessions." *Africa* 87 (4): 758–79.

Lyell, Charles. 1830. *Principles of geology, being an attempt to explain the former changes of the earth's surface by references to causes now in operation* (Vol. I). London: Murray.

Malkki, Liisa. 1992. "National Geographic: The Rooting of Peoples and the Territorialization of National Identity among Scholars and Refugees." *Cultural Anthropology* 7 (1): 24–44.

Mallon, Florencia E. 1994. "The Promise and Dilemma of Subaltern Studies: Perspectives from Latin American History." *American Historical Review* 99 (5): 1491–515.

Mallory, Ian A. 1990. "Conduct Unbecoming: The Collapse of the International Tin Agreement." *American University National Law Review* 5 (3): 835–92.

Mamani, Lidia. 2018. "El número de cooperativas mineras se duplicó en 12 años." *Página Siete*, January 26, 2018.

Mansfield, Becky. 2003. "Fish, Factory Trawlers, and Imitation Crab: The Nature of Quality in the Seafood Industry." *Journal of Rural Studies* 19 (1): 9–21.

Mantell, Charles L. 1949. *Tin: Its Mining, Production, Technology, and Applications*. New York: Reinhold.

Marchesi, Greta. 2014. "Polar Bears, Cactus, and Natives: Race, Agrarian Reform, and Environmental Determinism in Latin America (1920–1950)." *Journal of Historical Geography* 45 (C): 82–91.

Mariobo Moreno, Pedro. 2007. *El cooperativismo minero: Solución, engaño o solución?* La Paz: CEPROMIN.

Marston, Andrea. 2015. "Autonomy in a Post-neoliberal Era: Community Water Governance in Cochabamba, Bolivia." *Geoforum* 64:246–56.

Marston, Andrea. 2019. "Strata of the State: Resource Nationalism and Vertical Territory in Bolivia." *Political Geography* 74: 102040.

Marston, Andrea. 2021. "Of Flesh and Ore: Material Histories and Embodied Geologies." *Annals of the American Association of Geographers* 111 (7): 2078–95.

Marston, Andrea, and Matthew Himley. 2021. "Earth Politics: Territory and the Subterranean—Introduction to the Special Issue." *Political Geography* 88: 102407.

Marston, Andrea, and Amy Kennemore. 2019. "Extraction, Revolution, Plurinationalism: Rethinking Extractivism from Bolivia." *Latin American Perspectives* 46 (2): 141–60.

Martínez, María Elena. 2004. "The Black Blood of New Spain: *Limpieza de Sangre*, Racial Violence, and Gendered Power in Early Colonial Mexico." *William and Mary Quarterly* 61 (3): 479–520.

Marx, Karl. (1847) 2008. *The Poverty of Philosophy*. New York: Cosimo Classics.

Marx, Karl. (1852) 1963. *The Eighteenth Brumaire of Louis Bonaparte*. New York: International Publishers.

Marx, Karl. (1859) 1904. *A Contribution to the Critique of Political Economy*. New York: International Library Publishing Co.

Marx, Karl. (1867) 1990. *Capital*. Vol. 1. London: Penguin Classics.

Marx, Karl. (1973) 1993. *Grundrisse: Foundations of the Critique of Political Economy*. London: Penguin Classics.

Marx, Karl, and Friedrich Engels. (1846) 1998. *The German Ideology*. New York: Prometheus.

Marx, Karl, and Friedrich Engels. (1848) 1998. *The Communist Manifesto*. London: Verso.

Masters, Adrian. 2018. "A Thousand Invisible Architects: Vassals, the Petition and Response System, and the Creation of Spanish Imperial Caste Legislation." *Hispanic American Historical Review* 98 (3): 377–406.

Mbembe, Achille. 2001. *On the Postcolony*. Berkeley: University of California Press.

McClintock, Anne. 1995. *Imperial Leather: Race, Gender and Sexuality in the Colonial Contest*. New York: Routledge.

McKittrick, Katherine. 2006. *Demonic Grounds: Black Women and the Cartographies of Struggle*. Minneapolis: University of Minnesota Press.

McNeish, John A. 2013. "Extraction, Protest and Indigeneity in Bolivia: The TIPNIS Effect." *Latin American and Caribbean Ethnic Studies* 8 (2): 221–42.

McNelly, Angus. "The Uncertain Future of Bolivia's Movement Toward Socialism." *New Labor Forum* 30 (2): 80–89.

McSweeney, Kendra, David J. Wrathall, Erik A. Nielsen, and Zoe Pearson. 2018. "Grounding Traffic: The Cocaine Commodity Chain and Land Grabbing in Eastern Honduras." *Geoforum* 95:122–32.

Melo Zurita, María de Lourdes. 2019. "Holes, Subterranean Exploration and Affect in the Yucatan Peninsula." *Emotion, Space and Society* 32: 100584.

Melo Zurita, María de Lourdes. 2020. "Challenging Sub Terra Nullius: A Critical Underground Urbanism Project." *Australian Geographer* 51 (3): 269–82.

Melo Zurita, María de Lourdes, Paul George Munro, and Donna Houston. 2018. "Un-earthing the Subterranean Anthropocene." *Area* 50 (3): 298–305.

Mendez, Manuel, Manuel Prieto, and Milton Godoy. 2020. "Production of Subterranean Resources in the Atacama Desert: 19th and Early 20th Century Mining / Water Extraction in the Taltal District, Northern Chile." *Political Geography* 81: 102194.

Mendoza, Jaime. (1925) 2016. *El factor geográfico en la nacionalidad boliviana*. La Paz: Vicepresidencia del Estado Plurinacional de Bolivia.

Mendoza, Jaime. (1933) 2016. *El macizo boliviana*. La Paz: Vicepresidencia del Estado Plurinacional de Bolivia.

Mendoza L., Gunnar. (1971) 2002. "D'Orbigny en Bolivia." In *El naturalista francés Alcide d'Orbigny en la visión de los bolivianos*, edited by René Danilo Arze Aguirre, 211–34. Lima: Institut Français d'Etudes Andines; La Paz: Plural Editores.

Mendoza Torrico, Fernando, and Félix Patzi González. 1997. *Atlas de los ayllus del Norte de Potosí, territorio de los antiguos Charka*. Potosí: Comisión Europea— Delegación en Bolivia (Programa de Autodesarrollo Campesino).

Merrifield, Andrew. 2013. *Henri Lefebvre: A Critical Introduction*. New York: Routledge.

Mitchell, Timothy. 2002. *Rule of Experts: Egypt, Techno-Politics, Modernity*. Berkeley: University of California Press.

Mitchell, Timothy. 2011. *Carbon Democracy: Political Power in the Age of Oil*. New York: Verso.

Mitre, Antonio. 1993. *Bajo un cielo de estaño: Fulgor y ocaso del metal en Bolivia*. La Paz: Asociación Nacional de Mineros Medianos.

Mohanty, Chandra Talpade. 1988. "Under Western Eyes: Feminist Scholarship and Colonial Discourses." *Feminist Review* 30 (1): 61–88.

Molina, Fernando. 2011. *El pensamiento boliviano sobre los recursos naturales*. La Paz: Fundación Vicente Pazos Kanki.

Mollett, Sharlene. 2006. "Race and Natural Resource Conflicts in Honduras: The Miskito and Garifuna Struggle for Lasa Pulan." *Latin American Research Review* 41 (1): 76–101.

Montenegro, Carlos. (1944) 1984. *Nacionalismo y coloniaje*. La Paz: Librería Editorial "G.U.M."

Mumford, Jeremy Ravi. 2012. *Vertical Empire: The Struggle for Andean Space in the Sixteenth Century*. New Haven, CT: Yale University Press.

Mumford, Lewis. (1934) 2010. *Technics and Civilization*. Chicago: University of Chicago Press.

Murphy, Michelle. 2012. *Seizing the Means of Reproduction: Entanglements of Feminism, Health, and Technoscience*. Durham, NC: Duke University Press.

Murphy, Michelle. 2017. "Alterlife and Decolonial Chemical Relations." *Cultural Anthropology* 32 (4): 494–503.

Murra, John. 1972. "El 'control vertical' de un máximo de pisos ecológicos en la economía de las sociedades andinas." In *Visita de la Provincia de Leon de Huanuco en 1562*, edited by John Murra, 427–76. Huanuco: Universidad Nacional Hermilio Valdizán.

Nash, Jennifer C. 2019. *Black Feminism Reimagined: After Intersectionality*. Durham, NC: Duke University Press.

Nash, June. (1979) 1993. *We Eat the Mines and the Mines Eat Us: Dependency and Exploitation in Bolivian Tin Mines*. New York: Columbia University Press.

Neilson, David. 2018. "In-Itself for-Itself: Towards Second-Generation Neo-Marxist Class Theory." *Capital & Class* 42 (2): 273–95.

Nelson, Diane M. 1999. *A Finger in the Wound: Body Politics in Quincentennial Guatemala*. Berkeley: University of California Press.

Nelson, Diane M. 2003. "'The More You Kill the More You Will Live': The Maya, 'Race,' and Biopolitical Hopes for Peace in Guatemala." In *Race, Nature, and the Politics of Difference*, edited by Donald S. Moore, Jake Kosek, and Anand Pandian, 122–46. Durham, NC: Duke University Press.

Nichols, Robert. 2020. *Theft Is Property! Dispossession and Critical Theory*. Durham, NC: Duke University Press.

Nicolas, Vincent, and Pablo Quisbert Condori. 2014. *Pachakuti: El retorno de la nación*. La Paz: Fundación PIEB.

Nogales Vera, Neyer. 2009. "Factores de diferenciación social en el cooperativismo minero." Undergraduate thesis, Universidad Mayor de San Simón.

Oblasser, Angela, and Eduardo Chaparro Avila. 2008. *Estudio comparativo de la gestión de los pasivos ambientales mineros en Bolivia, Chile, Perú y Estados Unidos*. Santiago: CEPAL.

O'Connor, Alan. 1990. "The Miners' Radio Stations in Bolivia: A Culture of Resistance." *Journal of Communication* 40 (1): 102–10.

Órgano Electoral Plurinacional. 2016. *Estadísticas electorales: Referendo constitucional 2016*. La Paz: Órgano Electoral Plurinacional de Bolivia.

Oguz, Zeynep. 2021. "Cavernous Politics: Geopower, Territory, and the Kurdish Question in Turkey." *Political Geography* 85: 102331.

Ollman, Bertell. 2003. *Dance of the Dialectic: Steps in Marx's Method.* Chicago: University of Illinois Press.

Omi, Michael, and Howard Winant. (1986) 2015. *Racial Formation in the United States.* New York: Routledge.

Oporto Ordóñez, Luis. 2007. *Uncía y Llallagua: Empresa minera capitalista y estrategias de apropiación real del espacio (1900–1935).* La Paz: IFEA / Plural Editores.

Oporto Ordóñez, Luis. 2017. "Llallagua: La ciudad del Estaño." *Revista de la Biblioteca y Archivo Histórico de la Asamblea Legislativa Plurinacional* 11 (53): 94–100.

Orlove, Benjamin S. 1998. "Down to Earth: Race and Substance in the Andes." *Bulletin on Latin American Resources* 17 (2): 207–22.

Parikka, Jussi. 2015. *A Geology of Media.* Minneapolis: University of Minnesota Press.

Peck, Jamie, and Adam Tickell. 2002. "Neoliberalizing Space." *Antipode* 34 (3): 380–404.

Peluso, Nancy Lee. 2017. "Plantations and Mines: Resource Frontiers and the Politics of the Smallholder Slot." *Journal of Peasant Studies* 44 (4): 834–69.

Pérez, María Alejandra, and María de Lourdes Melo Zurita. 2020. "Underground Exploration beyond State Reach: Alternative Volumetric Territorial Projects in Venezuela, Cuba, and Mexico." *Political Geography* 79: 102144.

Perreault, Tom. 2005. "State Restructuring and the Scale Politics of Rural Water Governance in Bolivia." *Environment and Planning A: Economy and Space* 37 (2): 263–84.

Perreault, Tom. 2006. "From the Guerra del Agua to the Guerra del Gas: Resource Governance, Neoliberalism and Popular Protest in Bolivia." *Antipode* 38 (1): 150–72.

Perreault, Tom. 2013. "Dispossession by Accumulation? Mining, Water and the Nature of Enclosure on the Bolivian Altiplano." *Antipode* 45 (5): 1050–69.

Perreault, Tom. 2020. "Bolivia's High Stakes Lithium Gamble: The Renewable Energy Transition Must Ensure Social Justice across the Supply Chain, from Solar Panels and Electric Vehicles to the Lithium Extraction That Fuels Them." *NACLA Report on the Americas* 52 (2): 165–72.

Peters, Kimberley, Philip Steinberg, and Elaine Stratford. 2018. "Introduction." In *Territory beyond Terra,* edited by Kimberley Peters, Philip Steinberg, and Elaine Stratford, 1–13. London: Rowman and Littlefield.

Pijpers, Robert. 2014. "Crops and Carats: Exploring the Interconnectedness of Mining and Agriculture in Sub-Saharan Africa." *Futures* 62 (A): 32–39.

Pijpers, Robert, Esther van de Camp, Eleanor Fisher, Luciana Massaro, Jorge Calvimontes, Lorenzo D'Angelo, and Cristiano Lanzano. 2021. "Mining 'Waste.'" *Etnofoor* 33 (2): 13–40.

Platt, Tristan. 1982. *Estado boliviano y ayllu andino: Tierra y tributo en el Norte de Potosí.* Lima: Instituto de Estudios Peruanos.

Pollard, Kenneth Michael. 2016. "Silica, Silicosis, and Autoimmunity." *Frontiers in Immunology* 7: Article 97.

Poole, Deborah. 1997. *Vision, Race, and Modernity: A Visual Economy of the Andean Image*. Princeton, NJ: Princeton University Press.

Postero, Nancy. 2007. *Now We Are Citizens: Indigenous Politics in Postmulticultural Bolivia*. Stanford, CA: Stanford University Press.

Postero, Nancy. 2017. *The Indigenous State: Race, Politics, and Performance in Plurinational Bolivia*. Berkeley: University of California Press.

Postone, Moishe. 1979. *Time, Labor, and Social Domination: A Reinterpretation of Marx's Critical Method*. Cambridge: Cambridge University Press.

Poveda Ávila, Pablo. 2014. *Formas de producción de las cooperativas mineras en Bolivia*. La Paz: CEDLA.

Poveda Ávila, Pablo. 2021. "Explotación y comercialización de oro en Bolivia." *CEDLA: Reporte Anual de Industrias Extractivistas* 6:65–117.

Povinelli, Elizabeth A. 1998. "The State of Shame: Australian Multiculturalism and the Crisis of Indigenous Citizenship." *Critical Inquiry* 24 (2): 575–610.

Povinelli, Elizabeth A. 2011. *Economies of Abandonment: Social Belonging and Endurance in Late Liberalism*. Durham, NC: Duke University Press.

Povinelli, Elizabeth A. 2016. *Geontologies: A Requiem to Late Liberalism*. Durham, NC: Duke University Press.

Prada Alcoreza, Raúl. 2007. "Articulaciones de la complejidad: Estado plurinacional." Red Universidad Indígena Intercultural. http://www.reduii.org/cii/sites /default/files/field/doc/Estado%20plurinacional%20%20R%20Prada.pdf.

Pratt, Mary Louise. 1992. *Imperial Eyes: Travel Writing and Transculturation*. New York: Routledge.

Quastel, Noah. 2017. "Pashukanis at Mount Polley: Law, Eco-Social Relations and Commodity Forms." *Geoforum* 81:45–54.

Querejazú Calvo, Roberto. (1977) 1998. *Llallagua: Historia de una montaña*. La Paz: Libros Maravillosos.

Quesada Elias, Juan Isidoro. 1991. "Alcides d'Orbigny y el General Ballivián." *Presencia Literaria* 7 (21): 2–4.

Quispe-Agnoli, Rocío. 2011. "Taking Possession of the New World: Powerful Female Agency of Early Colonial Accounts of Perú." *Legacy: A Journal of American Women Writers* 28 (2): 257–89.

Radcliffe, Sarah, and Sallie Westwood. 1996. *Remaking the Nation: Place, Identity and Politics in Latin America*. London: Routledge.

Reader, John. 2008. *Potato: A History of the Propitious Esculent*. New Haven, CT: Yale University Press.

Redwood, Stewart D. 2003. "'The Father of Bolivian Geology': Friedrich (Federico) Ahlfeld (1892–1982)." *Mineralogical Record* 34 (3): 225–33.

Reinaga, Fauto. 1969. *La revolución india*. La Paz: Ediciones Partido Indio de Bolivia.

Richardson, Tanya, and Gissa Weszkalnys. 2014. "Introduction: Resource Materialities." *Anthropological Quarterly* 87 (1): 5–30.

Rifkin, Mark. 2019. *Fictions of Land and Flesh: Blackness, Indigeneity, Speculation*. Durham, NC: Duke University Press.

Riofrancos, Thea. 2020. *Resource Radicals: From Petro-Nationalism to Post-Extractivism in Ecuador.* Durham, NC: Duke University Press.

Rivera Cusicanqui, Silvia. 1987. *"Oppressed but Not Defeated": Peasant Struggles among the Aymara and Qhechwa in Bolivia, 1900–1980.* Geneva: United Nations Research Institute for Social Development.

Rivera Cusicanqui, Silvia. 1992. *Ayllus y proyectos de desarrollo en el norte de Potosí.* La Paz: THOA.

Rivera Cusicanqui, Silvia. 1997. "La noción de 'derecho' o las paradojas de la modernidad postcolonial: Indígenas y mujeres en Bolivia." *Temas Sociales*, no. 19, 27–52.

Rivera Cusicanqui, Silvia. 2003. "El mito de la pertenencia de Bolivia al 'mundo occidental': Requiem para un nacionalismo." *Temas Sociales*, no. 24, 64–100.

Rivera Cusicanqui, Silvia. 2004. "Reclaiming the Nation." *NACLA Report on the Americas* 38 (3): 19–23.

Rivera Cusicanqui, Silvia. 2015. "Violencia e interculturalidad: Paradojas de la etnicidad en la Bolivia de hoy." *Telar: Revista del Instituto Interdisciplinario de Estudios Latinoamericanos*, no. 15, 49–70.

Rivera Cusicanqui, Silvia. 2018. *Un mundo ch'ixi es possible: Ensayos desde un presente em crisis.* Buenos Aires: Tinta Limón.

Rivera Cusicanqui, Silvia. 2020. "Bolivia's Lesson in Triumphalism." In *Toward Freedom's Bolivia Reader: Voices on the Political and Social Crisis following the October 2019 Elections in Bolivia*, edited by Dawn Paley, 40–48. Burlington, VT: Toward Freedom. http://towardfreedom.org/wp-content/uploads/2020/01/Toward-Freedoms-Bolivia-Reader.pdf.

Robbins, Paul. 2012. *Lawn People: How Grasses, Weeds, and Chemicals Make Us Who We Are.* Philadelphia: Temple University Press.

Robinson, Cedric J. (1983) 2020. *Black Marxism: The Making of the Black Radical Tradition.* Chapel Hill: University of North Carolina Press.

Rodenbiker, Jesse. 2019. "Uneven Incorporation: Volumetric Transitions in Peri-urban China's Conservation Zones." *Geoforum* 104:234–43.

Rodríguez García, Huáscar. 2012. *La choledad antiestatal: El anarchosindicalismo en el movimiento obrero boliviano (1912–1965).* La Paz: Muela del Diablo Editores.

Ross, Corey. 2014. "The Tin Frontier: Mining, Empire, and Environment in Southeast Asia, 1870s–1930s." *Environmental History* 19 (3): 454–79.

Rudwick, Martin J. S. 1997. *Georges Cuvier, Fossil Bones, and Geological Catastrophes: New Translations and Interpretations of the Primary Texts.* Chicago: University of Chicago Press.

Ruiz Arrieta, Adriana Gloria. 2012. "Tramas de sentidos y significaciones durante las nacionalizaciones mineras de Huanuni y Colquiri en Bolivia." *Revista Latinoamerica de Antropología del Trabajo* 2 (3): 1–25.

Salaman, Redcliffe N. (1949) 1985. *The History and Social Influence of the Potato.* Cambridge: Cambridge University Press.

Sandoval, Chela. 1991. "U.S. Third World Feminism: The Theory and Method of Oppositional Consciousness in the Postmodern World." *Genders*, no. 10, 1–24.

Sanjinés C., Javier. 2004. *Mestizaje Upside-Down: Aesthetic Politics in Modern Bolivia*. Pittsburgh: University of Pittsburgh Press.

Schmitt, Carl. (1934) 1985. *Political Theology: Four Chapters on the Concept of Sovereignty*. Translated by George Schwab. Cambridge, MA: MIT Press.

Scott, Heidi V. 2008. "Colonialism, Landscape and the Subterranean." *Geography Compass* 2 (6): 1853–69.

Shanin, Teodor. 1982. "Defining Peasants: Conceptualisations and De-conceptualisations: Old and New in a Marxist Debate." *Sociological Review* 30 (3): 407–32.

Shever, Elana. 2012. *Resources for Reform: Oil and Neoliberalism in Argentina*. Stanford, CA: Stanford University Press.

Shultz, Jim. 2019. "Bolivia in Crisis: Don't Mistake a Public Uprising for a Coup." Medium, November 12, 2019. https://jimshultz716.medium.com/bolivia-in-crisis-4ef2f25471ed.

Silverblatt, Irene. 1987. *Moon, Sun, and Witches: Gender Ideologies and Class in Inca and Colonial Peru*. Princeton, NJ: Princeton University Press.

Simpson, Audra. 2014. *Mohawk Interruptus: Political Life across the Borders of Settler States*. Durham, NC: Duke University Press.

Smale, Robert L. 2010. *I Sweat the Flavor of Tin: Labor Activism in Early Twentieth-Century Bolivia*. Pittsburgh: University of Pittsburgh Press.

Smith, James H. 2022. *The Eyes of the World: Mining the Digital Age in the Eastern DR Congo*. Chicago: University of Chicago Press.

Smith, Neil. 1984. *Uneven Development: Nature, Capital, and the Production of Space*. Athens: University of Georgia Press.

Sneddon, Chris. 2007. "Nature's Materiality and the Circuitous Paths of Accumulation: Dispossession of Freshwater Fisheries in Cambodia." *Antipode* 39 (1): 167–93.

Sorrensen, Cynthia. 2014. "Making the Subterranean Visible: Security, Tunnels, and the United States–Mexico Border." *Geographical Review* 104 (3): 328–45.

Spedding, Alison. 2004. *Kawsachun coca: Economía campesina cocalera en los Yungas y el Chapare*. La Paz: PIEB.

Squire, Rachael, and Klaus Dodds. 2020. "Introduction to the Special Issue: Subterranean Geopolitics." *Geopolitics* 25 (1): 4–16.

Starn, Orin. 1991. "Missing the Revolution: Anthropologists and the War in Peru." *Cultural Anthropology* 6 (1): 63–91.

Starosielski, Nicole. 2015. *The Undersea Network*. Durham, NC: Duke University Press.

Steinberg, Philip, and Kimberley Peters. 2015. "Wet Ontologies, Fluid Spaces: Giving Depth to Volume through Oceanic Thinking." *Environment and Planning D: Society and Space* 33 (2): 247–64.

Stepan, Nancy Leys. 1991. *"The Hour of Eugenics": Race, Gender, and Nation in Latin America*. Ithaca, NY: Cornell University Press.

Stern, Steve J., ed. 1987. *Resistance, Rebellion, and Consciousness in the Andean Peasant World, 18th to 20th Centuries*. Madison: University of Wisconsin Press.

Stern, Steve J. 1988. "Feudalism, Capitalism, and the World-System in the Perspective of Latin America and the Caribbean." *American Historical Review* 93 (4): 829–72.

Stern, Steve J. 1993. *Peru's Indian Peoples and the Challenge of Spanish Conquest: Huamanga to 1640*. Madison: University of Wisconsin Press.

Stockwell, Alison Marie. 2015. "Capturing Vulnerability: Toward a Method for Assessing, Mitigating, and Monitoring Gendered Violence in Mining Communities in British Columbia." PhD diss., University of British Columbia.

Stoler, Ann Laura. 2013. "'The Rot Remains': From Ruins to Ruination." In *Imperial Debris: On Ruins and Ruination*, edited by Ann Laura Stoler, 1–35. Durham, NC: Duke University Press.

Svampa, Maristella. 2019. *Neo-extractivism in Latin America: Socio-environmental Conflicts, the Territorial Turn, and New Political Narratives*. Cambridge: Cambridge University Press.

Swyngedouw, Erik. 1999. "Modernity and Hybridity: Nature, Regeneracionismo, and the Production of the Spanish Waterscape, 1890–1930." *Annals of the Association of American Geographers* 89 (3): 443–65.

TallBear, Kim. 2013. *Native American DNA: Tribal Belonging and the False Promise of Genetic Science*. Minneapolis: University of Minnesota Press.

TallBear, Kim. 2017. "Beyond the Life / Not Life Binary: A Feminist-Indigenous Reading of Cryopreservation, Interspecies Thinking and the New Materialisms." In *Cryopolitics: Frozen Life in a Melting World*, edited by Joanna Radin and Emma Kowal, 179–202. Cambridge, MA: MIT Press.

Tandeter, Enrique. 1992. *Coacción y mercado: La minería de la plata en el Potosí colonial, 1692–1826*. Buenos Aires: Editorial Sudamericana.

Tassi, Nico. 2017. *The Native World-System: An Ethnography of Bolivian Aymara Traders in the Global Economy*. Oxford: Oxford University Press.

Taussig, Michael T. 1980. *The Devil and Commodity Fetishism in South America*. Chapel Hill: University of North Carolina Press.

Taylor, Keeanga-Yamahtta, ed. 2017. *How We Get Free: Black Feminism and the Combahee River Collective*. Chicago: Haymarket.

Taylor, Lucy. 2004. "Client-ship and Citizenship in Latin America." *Bulletin of Latin American Research* 23 (2): 213–27.

Thomas, Peter. 2009. *The Gramscian Moment: Philosophy, Hegemony and Marxism*. Leiden: Brill.

Thompson, Edward P. 1971. "The Moral Economy of the English Crowd in the Eighteenth Century." *Past and Present*, no. 50, 76–136.

Thomson, Sinclair. 2002. *We Alone Will Rule: Native Andean Politics in the Age of Insurgency*. Madison: University of Wisconsin Press.

Todd, Zoe. 2016. "An Indigenous Feminist's Take on the Ontological Turn: 'Ontology' Is Just Another Word for Colonialism." *Journal of Historical Sociology* 29 (1): 4–22.

Tubb, Daniel. 2020. *Shifting Livelihoods: Gold Mining and Subsistence in the Chocó, Colombia*. Seattle: University of Washington Press.

Tuck, Eve. 2014. "ANCSA as X-Mark: Surface and Subsurface Claims of the Alaska

Native Claims Settlement Act." *Alaska Native Studies in the 21st Century* 1 (1): 240–72.

US Geological Survey. 2013. *Metal Prices in the United States through 2010.* Reston, VA: US Geological Survey.

Valdivia, Gabriela. 2015. "Oil Frictions and the Subterranean Geopolitics of Energy Regionalisms." *Environment and Planning A: Economy and Space* 47 (7): 1422–39.

Van Buren, Mary. 1996. "Rethinking the Vertical Archipelago: Ethnicity, Exchange, and History in the South Central Andes." *American Anthropologist* 98 (2): 338–51.

Van Buren, Mary, and Barbara H. Mills. 2005. "Huayrachinas and Tocochimbos: Traditional Smelting Technology of the Southern Andes." *Latin American Antiquity* 16 (1): 3–25.

Van Cott, Donna Lee. 2000. "A Political Analysis of Legal Pluralism in Bolivia and Colombia." *Journal of Latin American Studies* 32 (1): 207–34.

Vasconcelos, José. (1925) 1979. *The Cosmic Race / La raza cósmica.* Los Angeles: California State University.

Verbrugge, Boris, and Beverly Besmanos. 2016. "Formalizing Artisanal and Small-Scale Mining: Whither the Workforce?" *Resources Policy* 47 (C): 134–41.

Wade, Peter. 1997. *Race and Ethnicity in Latin America.* London: Pluto.

Wade, Peter. 2017. *Degrees of Mixture, Degrees of Freedom: Genomics, Multiculturalism, and Race in Latin America.* Durham, NC: Duke University Press.

Wallace-Sanders, Kimberly. 2002. *Skin Deep, Spirit Strong: The Black Female Body in American Culture.* Minneapolis: University of Minnesota Press.

Walls, Laura Dassow. 2009. *The Passage to Cosmos: Alexander von Humboldt and the Shaping of America.* Chicago: University of Chicago Press.

Wanderley, Fernanda. 2017. "Entre el extractivismo y el vivir bien: Experiencias y desafíos desde Bolivia." *Estudios Críticos del Desarrollo* 7 (12): 211–47.

Wang, Chi-Mao. 2021. "Securing the Subterranean Volumes: Geometrics, Land Subsidence and the Materialities of Things." *Environment and Planning D: Society and Space* 39 (2): 218–36.

Watts, Michael J. 2001. "Petro-Violence: Community, Extraction, and Political Ecology of a Mythic Commodity." In *Violent Environments*, edited by Nancy Lee Peluso and Michael Watts, 189–212. Ithaca, NY: Cornell University Press.

Watts, Michael J. 2004. "Antinomies of Community: Some Thoughts on Geography, Resources and Empire." *Transactions of the Institute of British Geographers* 29 (2): 195–216.

Watts, Michael J. 2008. "Anatomy of an Oil Insurgency: Violence and Militants in the Niger Delta, Nigeria." In *Extractive Economies and Conflicts in the Global South: Multi-Regional Perspectives on Rentier Politics*, edited by Kenneth Omeje, 51–74. New York: Routledge.

Webber, Jeffery R. 2014. "Revolution against 'Progress': Neo-extractivism, the Compensatory State, and the TIPNIS Conflict in Bolivia." In *Crisis and Contra-*

diction: Marxist Perspectives on Latin America in the Global Political Economy, edited by Susan Spronk and Jeffery R. Webber, 302–33. Leiden: Brill.

Weismantel, Mary. 2001. *Cholas and Pishtacos: Stories of Race and Sex in the Andes.* Chicago: University of Chicago Press.

Weizman, Eyal. 2007. *Hollow Land: Israel's Architecture of Occupation.* London: Verso.

Williams, Raymond. 1980. "The Idea of Nature." In *Problems in Materialism and Culture: Selected Essays*, 67–85. London: Verso.

Wolfe, Patrick. 2006. "Settler Colonialism and the Elimination of the Native." *Journal of Genocide Research* 8 (4): 387–409.

Woon, Chih Yuan, and Klaus Dodds. 2021. "Subterranean Geopolitics: Designing, Digging, Excavating and Living." *Geoforum* 127:349–55.

Woon, Chih Yuan, and J. J. Zhang. 2021. "Subterranean Geopolitics, Affective Atmosphere and Peace: Negotiating China-Taiwan Relations in the Zhaishan Tunnel." *Geoforum* 127:390–400.

Wynter, Silvia. 2003. "Unsettling the Coloniality of Being/Power/Truth/ Freedom: Toward the Human, after Man, Its Overrepresentation—an Argument." CR: *The New Centennial Review* 3 (3): 257–337.

Yashar, Deborah J. 2005. *Contesting Citizenship in Latin America: The Rise of Indigenous Movements and the Postliberal Challenge.* Cambridge: Cambridge University Press.

Ybarra, Megan. 2018. *Green Wars: Conservation and Decolonization in the Maya Forest.* Berkeley: University of California Press.

Young, Kevin A. 2017. *Blood of the Earth: Resource Nationalism, Revolution, and Empire in Bolivia.* Austin: University of Texas Press.

Yusoff, Kathryn. 2017. "Geosocial Strata." *Theory, Culture and Society* 34 (2–3): 105–27.

Yusoff, Kathryn. 2018. *A Billion Black Anthropocenes or None.* Minneapolis: University of Minnesota Press.

Zambrana B., Amílcar. 2014. *El pueblo afroboliviano: Historia, cultura y economía.* La Paz: FUNPROEIB Andes.

Zavaleta Mercado, René. 1986. *Lo nacional-popular en Bolivia.* Mexico City: Siglo XXI Editores.

Zimmerer, Karl S. 2015. "Environmental Governance through 'Speaking Like an Indigenous State' and Respatializing Resources: Ethical Livelihood Concepts in Bolivia as Versatility or Verisimilitude?" *Geoforum* 64:314–24.

Bolivia: as Alto Peru, 16; Chinese influence in, 196; commodity boom (early 2000s), 6, 69, 116, 199, 226; constitution of 1995, 21; constitution of 2009, xii, xii–xiii, 4, 6, 32–33, 68, 195, 202, 230–31; developmentalist dynamic, 201–2; "dual republics" system, 16, 19; early colonial period (mid-1500s), 27, 34, 35, 36–40; early republican period (after 1825), 27, 34, 35, 40–50; independence from Spain, 16, 27, 32, 40, 107; *la media luna* (half-moon) departments, 200–201, 227; maps of, *23, 24*, 41; material history of, xvi, 14–15, 225–26; military coup of 1964, 55, 84; New Economic Policy, 20, 65; postrevolutionary period (after 1952), 27, 34, 35, 50–56; poverty rate, 6, 238n9; referendum of February 2014, 196–97, 203–4, 226, 227; right-wing groups, xiv, 7–8, 227; "State of '52," 19–20. *See also* National Revolution of 1952

boundaries: campesinos separated from miners, 28, 92, 94–95, 97; between classes, 72–73; clothing as signal for, 143–44; colonial generation of, 46; between cooperatives, 61; distinction between life and nonlife, xv, 124–25, 127–29; between hierarchies of workers, 111–13, 236; racial, 96, 111, 140; between state and mining cooperatives, 192, 205; surface as legally separate from subterranean, 33–34, 37, 49, 55–56, 58, 210, 230, 233; transgressions of, 8, 28, 95, 102, 117, 122, 140, 147, 161, 226

bourgeoisie, 3, 70, 133, 238–39n13, 245n6, 252n2

Braun, Bruce, 49

budles (buddle pits), 148–49, *149*, 153, 185

Burke, Melvin, 80

Butler, Judith, 136, 145

cabecillos (work crew leaders), 151, 210

Camacho, Luis Fernando, 227–28

campesinos (peasants; smallholding farmers), 72, 92, 94; dirt linked to by mestizo miners, 137–38, 161; military-campesino pact, 98; as miners, 84, 91–94, *93*, 101–3, 110–13, 137–38, 235; racialization of, 94–98; separation from miners, 28, 92, 94–95, 97; unions, 3, 19, 24, 98–100, 117, 121. *See also agro-mineros* (agricultural miners); CSUTCB (Confederación Sindical Única de Trabajadores Campesinos de Bolivia); Indigenous peoples

canal workers' union (*lameros and palliris*), 87, *88*

Cancañiri mine shaft (Llallagua), 59, 90, 182, 194

Canessa, Andrew, 97

capitalism, 13, 166–67, 250n2; abandonment by, 167; and dialectical relationship with workers, 9–10; geographic restructuring, 62–63, 244n2; spatial fix, 28, 62–63, 89, 242n2

capitulaciones (general mining rights), 38

Carnival, 191, 193–96

cassiterite (stannic oxide), 86, 148, 149, 161, 181

Catavi submunicipality (Llallagua), 75, 78, 171, 174–76

Catavi union, 87, *88*

Centenario (ex-company miners), 83

Central Potosí, regional federation of, 211–12

Chaco War (1932–35), 17–18, 78, 111, 141, 225, 240n34

ch'alla (blessing), 136, 146

chancadora (crusher), 183

Chile, 21–22, 100

China, 196

cholas, 98, 107, 140, 143, 251n13

chuño production, 106, 114

circulation of resources, 102, 106, 204

citizenship, 5, 19, 39, 143, 158, 232

class consciousness, 62, 70–73, 104

class formation, 86, 134–35

clothing, 143–45; of *cholas*, 98, 140, 143, 251n13; *de pollera*, 98, 107, 140, 143–44, 171; *de vestido*, 143–44; and racialization, 144–45, 161

COB (Central Obrera Boliviana), 117–18

coca cultivation, 66–67, 186, 245n7

cocalero (coca growers') unions, 21, 66, 245n7

Cochabamba Water War (1999), 21, 66

Código de Minería (1997), 31

colas arenas (sand tailings), 182

Cold War, 28, 85, 169

Colloredo-Mansfield, Rudi, 139, 143

colonialism: early colonial period (mid-1500s), 27, 34, 35, 36–40; education as tool of, 143; El Tío's racialized traits, 154; gendered division of space, 140; impact on Marx's thought, 105; indirect rule, 15; *k'ajcheo* labor category, 82, 235; mining associated with legacies of, 168; "pact of reciprocity," 16, 100; *patronaje*, 206–9; resource extraction, 29; settler, 96, 247–48n4. *See also* Spain

Colquechaca, tin workers from, 100

Colquiri (state-owned mine), 235

Columbus, Christopher, 37

COMIBOL (Corporación Minera de Bolivia), 19, 20, 26, 33; abandoned properties used by mining cooperatives, 67; context for, 78; creation of, 54–55, 78, 100; dismantling of (1985), 61, 86, 89, 190; and private-cooperative partnerships, 32; Sink and Float Plant as property of, 184; and subsidiary organizations, 81–85, 89; and Triangular Program, 79–81

commodity boom (early 2000s), 6, 69, 116, 199, 226

communitarian mining, 230–33

Compañía Estañífera de Llallagua (Tin Company of Llallagua), 100

Comunidad, indigenismo y marxismo (Ferreira), 72

CONAMAQ (Consejo Nacional de Ayllus y Markas del Qullasuyu), 7, 56, 98–99, 101, 117, 121, 252n6, 254n6; and communitarian mining, 230–31, 233; distancing from MAS, 202

concentration plants, 13, 87–88, 153, 180–85; small-scale operations, 182–84. *See also* Sink and Float Plant

consciousness, 10, 28, 132–34, 251n6; class consciousness, 62, 70–73, 104; labor's effect on, 10, 28, 70–73, 133–35; and mode of production, 71, 239n16; nature's shaping of, 11, 162, 169; political, 72, 104–6, 133, 141, 249n20. *See also* subject formation

Constitution of 1995, 21

Constitution of 2009 (Constitución Política del Estado de 2009; CPE), xii–xiii, 4, 6, 195; electoral representation system, 202; "productive actors" in, xiii, 33, 68, 230–31

cooperativistas (cooperative miners), xii; individualization of risks and profits, 3; as petit bourgeois entrepreneurs, 3, 133

copajira (acidic orange water), 128–29

Córdova, Héctor, 56–58

Cornwall, tin from, 74, 148

Coronil, Fernando, 200

Córrego do Feijão iron ore mine (Brazil), 146

"cosmic race," 17, 52

criollos (Creoles), 16, 19, 32

"cruel optimism," 184

CSUTCB (Confederación Sindical Única de Trabajadores Campesinos de Bolivia), 56, 98–99, 117, 121

cuadrículas (grids), 34, 151

cuadrillas (work crews), 101, 113–16, 122, 136–37

Cuvier, Georges, 42–46, 55, 242–43nn13–16

d'Avignon, Robyn, 82

Day of the Plurinational State (January 21), 217–19, 222

"dead labor," 151, 170, 252n1

debt, 153, 207–9

decrees (Decretos Supremos): DS 2888 (dynamite prohibited at protests), 214–15; DS 2891 (restriction of private-cooperative partnerships), 215; DS 3196, 54; DS 4721 (Indigenous participation in mining), 230; DS 21060, 61, 65–66, 72–73, 79, 85, 87; DS 21377, 67; of September 1, 2016, 215–16. *See also* laws

degeneration theory, 43, 52

degradation, 29, 126–31, 250n3; across the life/nonlife divide, 127–28; *agotada* as term for, 126–27; bodily, 130–31, 143; *despintar* (lose color, disappear), 123; formation linked to, 143; forms of, 128; geological, 28–29; *gradation*, as term, 127; as process of deterioration, 127–28; of rock, 139, 153; slide from endurance to exhaustion, 127; by water, 150. *See also* formation

Deister, Emil, 183

de la Cadena, Marisol, 190

DENAGEO (Departamento Nacional de Geología), 51, 55

De re metallica (Agricola), 148, *149*

desmontes (slag heaps), 76, 163–64, *164*, 179, *179*, *180*, 185–87, *186*; and private-cooperative partnerships, 215–16; as term, 164

diabolus metallorum (the devil's metal), 74, 246n14

dialectical materialism, 9

Díaz Romero, Belisario, 49

differentiation: bodies, differential mattering of, 10, 133, 136, 139, 159–62; production of, 12–13; waste, understanding of, 243–44n3

dirt, concept of, 137–39, 161, 251n9

discovery, doctrine of, 33

disease, racialization of, 138–39

dissection, 42–45, 53, 243n14; as metaphor for racialization, 44–45; visual, 47–48

Freiberg School of Mines, 46

FSTMB (Federación Sindical de Trabajadores Mineros de Bolivia), 78, 79, 118

future imaginaries, 165–66, 189; and imaginaries of invisible geological wealth, 190–91; leaders from Llallagua-Uncía memorialized, 191–92; nostalgia for, 167–68, 170, 191

galamseys (Ghana), 3, 63

García Linera, Álvaro, 202

garimpeiros (Brazil), 3

Gas War (2003), 21–22, 66, 227, 228

gaze, scientific, 42, 46, 48, 54–56

gender, 211, 252–53n7; and access to health care, 159; complementarity, 140; and industrial ruins, 165–66; and labor hierarchies, 125, 146–47, 161–62; and natural resources, xv, 52; and racialization, 140, 144–45, 155; space divided by, 140–42

GEOBOL (Servicio Geológico de Bolivia), 55

geognosy, 43, 46–47, 47, 242–43n13

geography, 12–14, 13, 52; physical geographers, 43, 46; and racialization, 52, 98. See also three-dimensional space

Geología de Bolivia (Ahlfeld), 53–54

geological knowledge, 27, 34; centralization of, 51, 54–55; environmental determinism and scientific racism linked, 35, 43–44, 243n16; geognosy, 43, 46–47; medical metaphor for, 50–51; mineralogy, 42; physical geography, 43, 46; in republican era, 40–50

geology: distinction between life and nonlife, xv, 124–25, 127–29; formation in, 28–29, 85–86, 126, 132–33, 161; separation of environmental from cultural data points, 49–50, 55

Godoy, Ricardo, 82

gold mining, 63, 69, 234–35, 245n8

Gramsci, Antonio, 133, 251n7

Granada, Battle of, 36, 37

Gran Chaco region, 17–18, 111, 200

Graulau, Jeannette, 37

Gudynas, Eduardo, 6, 226

Gustafson, Bret, 201

haciendas, 19, 97, 98, 100, 206

Haggard, Henry Rider, 152

hands, as indicative of labor, 142–45, 147, 148, 161–62

Harris, Olivia, 111

Hartman, Saidiya, 139

Harvey, David, 28, 62–63, 89

health issues: ad COVID-19 pandemic, 234; disease, racialization of, 138–39; dust, impact on, 156–61; injuries as badges of honor, 131; lung diseases, 126–27, 152; silicosis, 156–58; toxic gases, 12, 130, 155; unequal distribution of, 87, 116, 126, 148–49

Hegel, Georg Wilhelm Friedrich, 134

Historia del movimiento obrero boliviano 1933–1952 (Lora), xi, 70

historical materialism, 62, 90, 124, 133–35; mode of production, 70–71, 239n15; reworking of, 224–25; and socionatural forces, 10–11, 135. See also material histories

historical products, individuals and groups as, 10, 133–34

Hochschild, Moritz (Mauricio), 18, 75

holidays, 191; Carnival, 191, 193–96; Day of the Plurinational State (January 21), 217–19, 222; San Juan celebrations, 80, 177

Huallpa, Diego, 15

Huanuni tin mine, 78, 120–21, 132, 235, 249–50n24; downstream impacts of, 146

Huber, Matthew T., 82

Humboldt, Alexander von, 42–43, 44, 46–47, 242–43n13; Tableau physique des Andes et pays voisins, 47

hydrocarbons, 7, 199–201, 234, 238n8. See also natural gas sector

IDH (Impuesto Directo a los Hidrocarburos), 199–201

Illanes, Rodolfo, murder of, 1, 4, 8, 29, 31, 57, 198, 205–6, 212–13, 237n1. See also roadblocks

imperial debris, 165, 167

import substitution industrialization (ISI), 64–65

imprisonment: of Indigenous leaders, 197; of miners, 190, 192, 205–6, 211, 213–14, 231, 254n17

Incan Empire, 15, 16, 96–97, 140

indigeneity, 95–99; agriculture disaggregated from, 116–17; collective territorial control struggles, 96, 98; indígena, as term, 97; and location, 96–97; and race, 95–96

indigenistas ("indigenists"), 25–26

Indigenous peoples, xii–xiii; autonomy, 96, 99–100, 225–26, 231–33; communitarian mining, 230–33; enslavement of, 38–39; feminization and infantilization of, 155, 207; forced resettlement (reducciones), 15–16, 39; indio permitido (permitted Indian) and indio prohibido (prohibited Indian), 248n6;

pación Popular (Law 1551), 21; Ley General de Trabajo, 118; Ley INRA (Ley del Instituto Nacional de Reforma Agraria; Law 1715), 21; mining law bill of 2014, 31–32, 35, 56–58, 215–16; "natural law" theories, 35–40; property law, subterranean, 27, 33–36, 43, 50. *See also* decrees (Decretos Supremos)

layoffs, 20, 61, 66, 79–80, 220, 245n6

leadership, 141, 164, 165, 191; hands, importance of in debates, 142–45; roles for marginalized people, 29, 198, 202–3, 209, 221–22

Lechín, Juan, 78, 79

left, Latin American, 9, 227, 238n6; new, xiii–xiv, 4–6; "pink tide," xiii–xiv

legal frameworks: separate systems for indigenous and Spanish, 39; subterranean separated from surface, 33–34, 36, 37, 49, 55–56, 58, 210, 230, 233; *sub terra nullius*, 33, 49, 58, 113, 233. *See also* decrees (Decretos Supremos); laws

letrado (literate citizen subject), 143

liberalism, 40–41; late, 127, 130, 162

life and nonlife, distinction between: in geology, xv, 124–25, 127–29; in late liberalism, 162

limpieza de sangre (blood purity), 36, 98

lithium resources, 228–29

Llallagua (tin-mining town, Norte Potosí), 22, 24; ayllus in, 99; Cancañiri mine shaft, 59, 90; as company town, 75; domestic violence in, 153–54; etymology of, 102–3, 248n12; form of neoliberalism in, 85; massacre of San Juan (June 23, 1967), 80–81, 177, 220; mining cooperatives, 24–25; state abandonment of, 61, 167; statues in Siglo XX Plaza, 107, 108. *See also* Uncía (tin-mining town, Norte Potosí)

llampu (rocks with tin flecks; soft, spongy, weak), 146–47, 147, 153

local perspectives, 14, 63, 167, 171–76; visions for future, 183–84

locatarios ("tenants"), 82–83; 20 de Octubre group, 84, 86, 88, 89; Dolores group, 86, 88, 89

Lora, Guillermo, xi, 70, 118, 238–39n13. *See also* Trotskyism

lorismo, 9, 133, 238–39n13

Lyell, Charles, 42

Machacamarca (town, Norte Potosí), 76, 172

machinery, 179–87, 247n2; burial of in 1980s, 192; Cold War "technology transfers," 169;

as "dead labor," 151, 170, 252n1; drills, 107, 115, 150–55; electrical generators, 189; elevators (*jaula*), 189–90; sense of time produced by, 166; used in concentration plants, 179–83. *See also* industrial ruins

Makene, Madoshi, 82

mak'unkus, as term for migrants, 102

Malaysian tin mines, 75, 76, 78

maps, 22; of Bolivia, 23, 24, 41; geological, 47, 51–55

Marchant Y., Victor E., 41–42

Marx, Karl, 10, 28; *A Contribution to the Critique of Political Economy*, 71; critiques of, 134–35; *The Eighteenth Brumaire of Louis-Napoleon Bonaparte*, 104, 246n1; *The German Ideology*, 134; on labor and consciousness, 133–34, 239n16; peasants, view of, 104–5; *The Poverty of Philosophy*, 169–70, 245–46n1; *Capital*, 71, 133, 239n14

Marxism, 9, 18, 71–72, 131; cooperative miners' rejection of, 133–35. *See also* Trotskyism

MAS (Movimiento al Socialismo), xiii, 72, 120, 172, 194, 197, 254n6; and cooperative mining federation support, 209–10; and ouster of Evo, 228–29; and Panduro roadblock, 212, 214, 217–19; and redistribution, 199–205; as "social movement state," 201. *See also* Morales, Evo

masacre blanca (white massacre; layoffs), 66

masculinity: domestic violence as performance of, 153–54; and drills, 151–53; formation of in mines, 141–43; mestizo, 107, 121, 125, 131, 155; silicosis diagnosis as marker of, 157; whiteness, relationship with, 25, 153. *See also* national inheritance (*patrimonio*)

massacre of San Juan (June 23, 1967), 80–81, 177, 178

material fixes: and economic crises, 62–64, 81–82, 88–89; nature, space, and technology, 88–89; and neoliberalization, 85–88; postrevolutionary, 78–85; and superfluous employees, 80, 83; and three-dimensional space, 28, 63, 89; in twentieth century, 73–78

material histories, 56, 89, 161, 197; of Bolivia, xvi, 14–15, 225–26; as both biophysical and sociocultural, 125; in *The German Ideology*, 134; as material geography, 89; of nature, xvi, 30; as place-based, 10, 12; of resource nationalism, 34, 95; of tin, 62, 95, 125. *See also* historical materialism

Sepúlveda, Juan Ginés de, 38–39
sex, as material attribute, 136
sexuality, violent, 25, 153
"shadow sovereigns," 64
Shever, Elana, 64, 68
Sierra Leone, 64
Siglo XX (mine shaft), 75, 127, 181
Siglo XX (submunicipality), 171, 176–77
Siglo XX (workers' union), 77–78, 83, *88*
Siglo XX plaza (Llallagua), 107, *108*, 177
silicosis, 156–58
silver mining, 15, 39, 73–74, 168; *k'ajcheo* practice, 81; miners' move from to tin industry, 100; patio process, 15–16; post-independence, 18
Sink and Float Plant, 76, 83–84, 87, 148–49, 179, 179–84; wastewater operations, 82–83, 87–88. *See also* concentration plants
slag heaps. *See desmontes* (slag heaps)
slavery, 96
"slime concentrators," 148
smallholders, 19, 105
social stratifications, 61–62; and architectural differences, 177; exhaustion, unequal distribution of, 127, 148; health outcomes, unequal distribution of, 87, 116, 126; impact of mineralogical variation on, 86–88, 102, 116, 125–26, 161; in *K'epirina* comic book, 91–94; labor hierarchies, 28–29, 63, 68, 86, 102, 236; material differences, 75, 161; and nature, 134–35; as norm, 69; spatial divisions, 13, 39, 49, 87–89, 100, 102, 177; vertical, 13, 63, 89, 114, 116, 230, 239n21
social structure, and "technological determinism," 169–70
socionatural forces, 10, 11, 13, 88
sociotechnical imaginary, 167, 190–91
sovereignty, 5; and communitarian mining, 232–33; Indigenous, 96, 98, 116–77, 130, 233; over subterranean, 14, 230; "shadow sovereigns," 64; theological interpretations of, 36–37
Soviet Union, 169
Spain, 15–16; Bolivia's independence from, 16, 27, 32, 40, 107; colonization of geographical areas, 96–97; ethical debates, 38–39, 242n6; founded through religious conquest, 36–37; one-fifth tax, 38, 39; religious education mandated by, 168; taxation systems, 37–39; Valladolid Debate of 1550, 38–39, 242n6; vertically stacked domains, 37, 240n28. *See also* colonialism

Spanish literacy, 143, 144
spatial fix, 28, 62–63, 89, 242n2. *See also* material fixes
spatialization: gendered division of space, 140–42; of legal frameworks in early colonial period, 39; vertical axis, 8. *See also* social stratifications; three-dimensional space; vertical spaces
Spencer, Herbert, 169
stannite, 86
Starn, Orin, 6
state: abandonment by, 61, 67, 130, 167, 199, 237n2; dismantling of institutions, 61, 65; as father figure, 209; as landlord, 210; as *patrón*, 35, 53, 58, 206–10; and rent seeking, 209, 215. *See also* nation
statues of miners, 107, *108*, 109, 110
steel, "tin-free," 79
Stoler, Ann, 165, 167
structuralist interpretations, 64, 71–72
subject formation, 28–29, 133–34, 161, 221; labor sites as sites of, 124–25, 197–98. *See also* consciousness; formation
subsidiary organizations, 81–85, 102, 246–47n20; *veneristas*, *lameros*, and *locatarios*, 82–84
subterranean: as archive, 8, 14; as full-body experience, 128–29; and global inequality, 13–14; "indigenization" of, 28, 91, 94, 122; land as separate from, 33, 49–50; local experiences of, 14; measurement in *cuadrículas* (grids), 34; as national inheritance (*patrimonio*), xii, 5, 14, 29, 33–35, 53, 57, 111, 197; and nationalism, xv, 5, 111; natural gas sector, 6, 199–200, 204, 212, 238n8; as natural resource, 51, 55; non-biological additions to, 129, 191; as ordered set of strata, 27, 35, 53; political impact of, 5, 8–14, 29, 198, 223–30; politically neutral construction of, 50–51, 56; and property law, 27, 33–36, 43, 50; as protagonist, 5; racial ambivalence of, 108; redistribution of, 199–205; retained in bodies, 156–61, 159, 198; and sovereignty, 14, 230; surface as legally separate from, 33–34, 37, 49–50, 55–56, 58, 210, 230, 233; as three-dimensional space, 13–14; underground as place to hide, 213; water and mining, 145–50
sub terra nullius, 33, 49, 58, 113, 233
Sucre, Antonio José de, 52
Sur Atocha federation, 211–12
surface: and feudalist model of encomiendas, 37; Indigenous peoples relegated to, 38, 40,

45, 45–49, 113, 117, 230; as legally separate from subterranean, 33–34, 37, 49–50, 55–56, 58, 210, 230, 233; potatoes relegated to, 106. *See also* vertical spaces

Tanzania, 82

Taussig, Michael, 154–55

taxation: cooperative miners' opposition to, 57–58, 199, 225; early colonial period, 37–38; forced labor, 38, 96, 97, 206; IDH (Impuesto Directo a los Hidrocarburos), 199–201; *Indian*, as tax category, 39; one-fifth tax, 38, 39; tribute, extraction of, 15, 16, 96, 100

technological determinism, 169–70

technological investments, 116, 184–87

"technology transfers," 169

teleférico (gondola-based public transit system), 180

temporality, 54, 165–66. *See also* future imaginaries; nostalgia; possibility

tercerización (outsourcing), 69

terra nullius, 33

Tesis de Pulacayo (Lora), 70

theft: by elites and foreign companies, 121, 249–50n24; as individualistic, 116; *jukeo* (ore theft), 30, 80–81, 189, 234, 246n17; *k'ajcheo* practice as, 81–82, 235; limitless term proposal as, 57; of *patria*, xi–xii, 35, 56–58; proliferation of thieves, 233–36; and property, 230; takeovers (*avasallamientos*), 234–36

theology, 27, 51, 58; *limpieza de sangre* (blood purity), 36, 98; "natural law" theories, 35–40; secularization of, 35, 36, 40–41, 58; sovereignty, interpretations of, 36–37

Thesis of Pulacayo, 70, 118

THOA (Taller de Historia Oral Andina), 20

three-dimensional space, 13–14; and material fix, 28, 63, 89; and social stratification, 63; and spatial incarceration of Indigenous peoples, 49. *See also* mineralogical variation

tierras comunitarias de origen (communal lands of origin), 21

tin: as alloying agent, 74; commodity substitutes for, 79; delisted from London Stock Exchange, 85; as *diabolus metallorum* (the devil's metal), 74, 246n14; as element Sn (metallic commodity), 62; expectations of wealth and progress linked with, 125; geological and chemical properties of, 28; grades of, 86, 130, 147; as key metal of industrial modernity, 109–10; material history of, 62, 95, 125; mineralogical variation in, 28, 85–86, 125–26; ore, geophysical characteristics of, 62, 125; ore types, 86, 126; recycling campaigns, 79; *roca sana* (healthy rock), 153; uses of, 74. *See also* tin-mining sector; tin ore

tin barons (*los barones del estaño*) 18, 75

tin-mining sector, xv; decline of, 18, 20, 130; denationalization of, 55; global tin cartel, 76–78; historical context, 60–62; international market, post–World War II, 79; market disruption (1985), 85, 166, 175, 199; and mestizo nationalism, 107–13; mestizos produced by, 107–8; military presence at, 180; mines as crucibles of nationalism, 111; national development bankrolled by, 166; nationalization of (1952), 3, 51, 54, 62, 65–66, 78, 90–91, 130; social stratification of, 61–62; Soviet offer of smelter, 169; symbolic importance of, 65–66; technological changes, 76; unfairness of extractive regime, 18; wastewater operations, 82–84, 87–88; waves of workers, 100

tin ore, 28; bauxite, 79; cassiterite (stannic oxide), 86, 148, 149, 161, 181; *colas arenas* (sand tailings), 182; in *desmontes* (slag heaps), 76, 163–64, *164*, 179, *179*, *180*, 185–87, *186*; differentiated quality of, 77, 82–88, 126, 147–48, 151; geophysical characteristics of, 62; hereditary rights to veins of, 115–16; *jukeo* (ore theft), 30, 80–81, 189, 234, 246n17; *mamo* ("sucked" ore), 149, 153, 183; as mestizo substance, 95; ore bodies, 153, 162; ore sands, 82–83, 87–88, 148; Patiño's discovery of, 73–75; pyrite, 86, 181, 183; tin sulfide, 124, 148, 149, 161

Tío, El (devil figure of mines), 151, 154–55, 194, 245n10, 246n14

TIPNIS (Territorio Indígena y Parque Nacional Isiboro Sécure), 7, 202, 231

tope (limit point of extraction), 115, 117

Torres, Juan José, 84

toxins: acidic water, 146–50, 161; gases, 12, 130, 155; and Indigenous sovereignty, 130

triad of state (*patron*), nation (*patria*), and subterranean (*patrimonio*), 35

Triangular Plan, 79–81, 83–84

Trotskyism, 9, 18, 70–72, 106, 131, 238n13, 246n13. *See also* Lora, Guillermo; Marxism

Tuck, Eve, 33

Túpac Katari Revolutionary Liberation Movement, 21

Uncía (tin-mining town, Norte Potosí), 22, 24, 171–74; ayllus in, 99; Centenario group, 83; as company town, 75; form of neoliberalism in, 85; mining cooperatives, 24–25; mining museum, 173; processing plants, 174; and railroad, 76, 77; state abandonment of, 61, 167. *See also* Llallagua (tin-mining town, Norte Potosí)

unions, 18–19, 253–54n15; anarcho-syndicalism, artisan-led, 18; autodidacts in, 131; campesino, 3, 19, 24, 98–100, 117, 121; *cocalero* (coca growers), 21, 66, 245n7; and denationalization of tin mining, 55; formation, interpretation of, 131–32; great leaders as miners, 141; imposition of by NGO, 101; mine workers,' and COB, 117–18; national unity, commitment to, 112–13; neoliberal dismantling of, xii, 64, 101, 131; in Norte, 99–102; and relocalized miners, 66–68, 76–77; Siglo XX, 77–78, 83, 88; Siglo XX plaza building, 177; and subsidiary organizations, 82. *See also* National Revolution of 1952

United States: Alliance for Progress, 79; antiunion collaboration with Bolivia, 55; Keynesian economics in, 64; settler colonialism, 96, 247–48n4; Strategic and Critical Materials Stockpiling Act (1939), 78; tin's association with modernity, 109–10; tin stockpile, 78, 85; and Triangular Plan, 79–81, 83–84, 169; US Operation Mission to Bolivia (USOM), 244n22; war on drugs, 66–67

UNSXX (Universidad Nacional Siglo XX), 70, 131–32, 151

USAID (United States Agency for International Development), 202

US Geological Survey, 54, 55

Valladolid Debate of 1550, 38–39, 242n6

Vasconcelos, José, 17, 52

veneristas, 82–83, 86–89, 88

Venezuela, 200

vertical spaces, 8–14, 239nn21–22, 240n28; ayllus as, 15; history as, 54; Spanish territories as stacked domains, 37; spatialization of social collectives, 3, 13, 89, 114, 116, 230, 239n21; state territorialization of, 55; variety within metalliferous veins, 85–86, 116. *See also* surface

Vice Ministry of Mining Cooperatives, 26

Villarroel, Walter, 120–21, 209

violence: domestic, 153–54; sexual, 25, 153

violence, political, 29, 197–98, 205; dynamite prohibited in protests (DS 2888), 214–15; legal retribution, 214–15; protest and punishment, 212–19; underground spaces for hiding from, 213

War of the Pacific (1879–83), 21–22

waste, differential understanding of, 243–44n3

wastewater operations, 82–83, 87–88

water, 145–50; acidic, 146–50, 161, 181; and *budles*, 148–49, *149*; *copajira* (acidic orange water), 128–29; and *llampu* selection, 146–47; and silicosis, 159

water supply, 21, 25

watía feast, 106

Weismantel, Mary, 97, 145, 153

Werner, Abraham Gottlob, 242–43n13

West Africa, 82

white supremacy, 25

Winant, Howard, 133

women: Aymara merchants, 119; *bien formada*, 143; white feminine vulnerability, myth of, 25

women miners, 139–41; *budle* operators, 148–49, *149*; cooperative, 60; drill operation by, 151; gendered labor division, 125, 146–47, 161–62; *llampu* selection by, 146–47; lung problems due to dust, 159–61; *socias cooperativistas mineras*, 119, 142; technological optimism of, 185; veins discovered by, 223; widows' inheritance of veins, 115, 160. *See also palliris* (women sorters)

workers: dialectical relationship with capitalists, 9–10; "informal," 118, 119, 234; *obreros*, 92, 94–96. *See also* miners; salaried miners; women miners

working class, European, 105

World War I, 76, 78

World War II, 78, 109

xanthate, 76, 87, 149, 181, 183

Young, Kevin A., xv, 18, 111, 243n22

Yusoff, Kathryn, 124, 125, 127

Zapata, Gabriela, 196, 252n1

Zapata scandal, 196–97, 203, 252n1

Zavaleta Mercado, René, 17–18, 246n13

Zurita, Maria de Lourdes Melo, 33

www.ingramcontent.com/pod-product-compliance
Lightning Source LLC
Chambersburg PA
CBHW071732270326
41928CB00013B/2643